THE INNOVATIVE YACHT

HOW TO IMPROVE A BOAT FOR COMFORT, CONVENIENCE AND PERFORMANCE

Andrew Simpson

WATERLINE

Published by Waterline Books
an imprint of Airlife Publishing Ltd
101 Longden Rd, Shrewsbury, England

© Andrew Simpson 1995

All rights reserved. No part of this publication may be reproduced, stored in a retrieval system or transmitted in any form or by any means, electronic, mechanical, photocopying, recording or otherwise, without the prior permission in writing of Waterline Books.

ISBN 1 85310 485 X

A Sheerstrake production.

A CIP catalogue record of this book is available from the British Library

Typeset by Servis Filmsetting Ltd, Manchester
Printed in England by Butler & Tanner, Frome and London

DEDICATION & THANKS

My thanks go to George Hayes, whose generosity of spirit allowed both *Fairlight* and his own lithe form to grace the cover. And also I must make a grateful nod towards Phaon Reid who, rudderless and marooned, was kind enough to proof-read the manuscript.

But, above all, this book is dedicated to Chele – my sailing companion, friend, and partner in all things.

Contents

Introduction	7
1 – The Shape of Things To Come	10
Identifying Function 11	
2 – Designed to Sail	15
Performance 15: Resistance 18: Displacement 20	
Prismatic Coefficient 22: Handling 24: Stability 29: Water Ballast 31	
3 – Aloft	33
Sails 33: Rigs 38: Reefing 45: Full-length Battens 50: Lazy Jacks 51	
Spinnakers 51: Cruising Chutes 53: High Performance Asymmetric Spinnakers 53	
Storm Canvas 54: Spars and Rigging 55	
4 – On Deck	57
Cockpit 57: Wheel or Tiller? 60: Fore and Side Decks 60: Winches 62	
Anchors, Warps and Windlasses 64: Stepping Ashore 66	
5 – Come Rain or Shine	68
Wheelhouses 69: Deck Saloons 70: The Dog House 71	
Windscreens 72: Spray Hoods and Dodgers 72: The Bimini 73	
Deck Awnings 74: Windscoops 76: Insect Screens 77	
6 – Self Steering	78
Windvane Gears 79: Electronic Auto-pilots 86	
7 – The Boat as a Habitat	89
Of Bunks and Things 94: Crew Size 95	
8 – Feeding the Five Thousand	98
Cookers 100: Refrigeration 105	

9 – Plumbing 114
Water Filters 115: Desalinators 115: Water Heaters 119
Showers 121: The Heads 122

10 – Blow Hot or Cold 125
Cabin Heaters 125: Air Conditioning 129

11 – Engines 130
Engine Size 130: Noise 132

12 – Electrics 139
How Much Electricity Do You Need? 140: Batteries 143
Engine Charging 145: Wind Generators 147: Solar Panels 149

13 – Electronics 154
Interface or Stand-Alone 156: Depth-Sounders 158
Electronic Logs 160: Wind Indicators 161: Electronic (Fluxgate) Compasses 161
Position Fixing 162: Chart Plotters 167: Radar 169

14 – Communications 171
VHF 171: Single Side-Band (SSB) 172: Amateur (Ham) Radio 172
Marine Telex 173: EPIRB 173: Satellite Communications 173
Weather Facsimile 174: Navtex 175

15 – Clothing 178

Index 184

Introduction

It was off Trafalgar that I met my Waterloo. It was there that the seeds of this book took root. For if there was one pivotal moment when I came to realise that there had to be an easier way, then that was probably it.

Being at that time short of cash, I had misguidedly agreed to deliver an ugly and otherwise awful catamaran to the Mediterranean – awful, I hasten to add, not because it was a catamaran, of which there are many fine examples – but because it was a wretched boat by any terms of reference.

At first the beast kept its vices to itself. Alongside it was a model of decorum. With more double berths than a honeymoon hotel, the accommodation was astonishing for a craft of that size. 'I think of it as a floating country cottage', the owner told me fondly. 'And soon it'll be my little casa in the sun.'

All of which was a fair description. As cottage or casa it performed acceptably. As a sailing machine it was lamentable.

We left Poole in late June. In a windward slog down the Channel, the catamaran pitched and pounded like a maddened horse. Beneath the bridge-deck, the water boiled up in a tumbling, resistant wall. Leeway was prodigious. Speed was slow: less than three knots we averaged, a lot of it sideways. A pitiful sail-plan, extravagant windage, and a portly hull form all did what they could to retard our progress towards Ushant. But, finally, after three exasperating days, we were swept on the tide through the Chenal du Four and were into the Bay of Biscay, jubilant and relieved.

Whereupon the steering died. Inspection revealed that the linkage had disintegrated. The damage appeared terminal. Onboard repairs were obviously out of the question, so, with steering lines jury rigged across the cockpit we took advantage of a now fair wind and laid a course for La Coruña.

Although the absolute calm that then invaded the area didn't really surprise us, the revelation that the outboard auxiliary would only run for ten minutes before overheating did. Now thwarted of meaningful progress under power, we settled back to sit it out. Eating, sleeping and backgammon formed essential elements in our routine.

By the sixth day we were bored but well rested, and it seemed almost an affront to our enforced lethargy to discover that a flexible tank had ruptured and had playfully dumped most of our drinking water into the bilge. Luckily the beer held out and we made it into La Coruña on day twelve.

When contacted from there, the owner expressed what we thought to be rather churlish irritation at the damage to his boat, and went on to make mumbling allusions to his personal poverty before refusing to provide replacement parts. 'You broke it, you fix it', was the general drift of his attitude. By now we had extracted the steering mechanism from the pedestal, and had unthreaded the cables that led back to the twin rudders. It was immediately obvious that the whole contrivance had been poorly conceived and had been chosen by the builders on the grounds of economy rather than adequacy.

For some reason we never thought to quit. An element of bloody-mindedness had entered into our relationship with the boat. We and it were now established adversaries and we, for our part, were not about to admit defeat. Between rejuvenating excursions to a nearby bar we were able to make some worthwhile refinements to our jury rig. The tank was patched and some emergency containers also put aboard. Soon we were ready to do battle again.

Proceeding south down the Iberian coast, the boat continued to keep us entertained. A breaking sea filled the dinghy and plucked one of the davits out of the aft deck. We broached spectacularly off the Islas Berlengas, driving the port hull into the trough of the wave, and bringing us as close to capsize as I ever wish to be. Then the electrics failed completely as we rounded Cape St. Vincent, leaving us dark and dumb amidst the milling traffic.

Finally, off Cape Trafalgar, the Levanter hit. And, from then on the trip went really badly.

So, was this voyage typical of what we might expect? Well, yes and no. I have experienced worse but also considerably better. Seven years later I was in the same area in another yacht and another gale. But there coincidence ends. This time we were sailing an intelligently designed thirty-three footer, well maintained and properly equipped. Unlike the catamaran of late and harrowing memory, this boat performed impeccably – as of course it should.

For sailing is about pleasure, not pain. Although rarely quantified, most yachtsmen keep a sort of internal balance-sheet, setting the good against the bad and reviewing on an ongoing basis whether it is all really worth it. Some become so discouraged they give up altogether. Others plod on with their passion, absorbing the inevitable privations, whilst wringing what enjoyment they can from those blissful times between the soggy bits. Of course the sea will always be a demanding place, often uncomfortable and always potentially dangerous. But there is a great deal that we can do to make life easier for ourselves. A well designed boat, chosen with care and fitted out with the right equipment for the job, will go a very long way towards keeping us comfortable and safe.

Introduction

But, before we proceed, let me disavow total objectivity. I have done my best, but this, I confess, is an opinionated book which may jar with the beliefs of others. There are many ways to cross an ocean and this is mine. To those who don't agree I give an understanding nod but no apology. There is room on the seas for all of us.

In the text that follows, all miles are nautical miles and, unless specified otherwise, boat lengths are Length Overall (LOA). Gallons are US gallons, but pints of beer remain pints of beer.

The author's yacht *Spook*, which he designed for his personal cruising requirements, contains many novel features.

Chapter 1
The Shape of Things To Come

It was the innovative Swiss architect Le Corbusier (1886 – 1965) who coined the oft-repeated design adage that 'form should follow function'. He was at that time designing private dwellings which he called *machines à habiter* (machines to live in), a term intended to reflect his admiration for the clean precision of machinery, not a suggestion that living in them should be an austere and mechanised existence. It was his view that the first purpose of houses was, quite obviously, to house people, and that every other consideration was secondary. He believed that if architects kept that simple truth in mind, their designs would evolve naturally, and have a pure beauty which no amount of studiously contrived styling could achieve. His own work remains a stunning testament to that ideal.

And, if the Le Corbusier message is true for buildings, how much more so must it be for boats? After all, the functions required of a house – protection from the weather, warmth, comfort, and efficiency – are all rather easily provided and, anyway, are similarly required of a sailing yacht. But, quite literally, a yacht must also be a complex and subtle machine – a *machine à naviguer* indeed – fashioned to live with the sterner demands of the environment in which it operates. It must move through both wind and water with optimum efficiency – maintaining that efficiency through a wide range of conditions; it must be easily navigable; it must be durable in its construction; and, above all, it must provide pleasure for its crew and keep them safe.

So, does the typical modern boat conform with the 'form following function' ideal? In many cases, unfortunately not.

As we know it today, sailing is a subject which owes as much to tradition as it does to technology. Although we might think that designs are always conceived in the clear, rational light of essential principles, the truth is often otherwise. And, perhaps more significantly, our individual perception of where exactly that truth lies can be blurred by our own preconceptions.

The Shape of Things to Come

Sailing is steeped in lore. We are a maritime people whose bond with the sea reaches back over many centuries. Whether we like it or not, our instincts – and therefore our tastes – are moulded by the past. For instance, you only have to look at the many replica 'traditional' boats around. Invariably, these are based on fishing boats; or, more precisely, on representations of fanciful fishing boat shapes. Of course, in their time, all working boats were wonderful examples of functional design – they would have been discarded had that not been the case. Each area had developed specific answers to local problems; the East Coast bawleys were as perfectly adapted to their own trade and waters as the Brixham trawlers and the Chesapeake Bay skipjacks were to theirs. And, distinctive though each type was, they had a lot in common: their keels were long and straight so they could dry out alongside to discharge their catch; their freeboards dipped towards the stern so nets or pots could more easily be hauled; they would lie a'hull with complete docility whilst their crews toiled on deck; and, most importantly as they were commercial craft, they were cheap to build and very robust. But they were never yachts; not then and not now.

Sentiment sells; the marine industry has long recognised this. The latter day replica, in all its counterfeit plastic gloss, is only the most conspicuous example of how our past still threads our thinking. Other, subtler myths persist.

Recently I was in conversation with a friend. He is the owner of a heavy but unquestionably able twenty-eight footer which he cruises extensively. At that time I was designing a 35′ (10.7m) sloop of approximately the same displacement as his own. Provocatively, I observed that this was really the same amount of boat more thinly spread – with some very worthwhile gains in performance and accommodation.

He seemed shocked, as if I had uttered some damnable heresy, and took time and a hefty swig of beer to compose himself.

'Yes, but I like the feel of a decent cruising boat beneath my feet', he growled at last.

So there you have it: another echo from astern. The word 'decent' in his mind was synonymous with heavy or solid, as if these were fundamental virtues to be revered. But traditional forms of construction inevitably produced heavy boats. This was more a limitation in technology, than conscious choice. Surely, now, modern techniques can liberate us from such historic constraints?

Identifying Function

Before we move on, it might be worth harkening back to the creed according to Le Corbusier. If form is to follow function then we must, of course, determine what that function is. Sailors and their ambitions come in all shapes and sizes. There is no standard specification, and certainly no right or wrong way to go sailing. But, before we can choose what sort of boat would suit us – the most vital decision of all – we must first define 'us'.

Here are some arbitrary categories:

Racers

The go-fast boys, active at the most competitive levels. Their boats are quick but often cranky and difficult to control. Although sheer speed is obviously vital, to be advantaged under the rating rules is also highly desirable. The quest for these advantages can result in grotesque hull shapes. Construction and equipment are both expensive and exotic. Racing crews care little for personal comfort, so the internal arrangements are usually Spartan bordering on squalid. Their boats can rapidly become obsolete and must therefore be considered almost disposable.

And yet, in some respects, that is yesterday's story. The waning popularity of the International Offshore Racing rule (IOR), and the increased enthusiasm for Channel Handicap and the International Measurement System (IMS), both of which are less mutilating in their influence, has nudged the trend towards more sensible boats which, without compromising speed, could also be successfully cruised.

Racer/Cruisers

The club racing fraternity who also like to gad around a bit. Performance is still important to them, but this is tempered by the odd lapse into more civilised behaviour. Their boats are usually up-rated versions of the faster cruising designs, or pensioned off racers from the higher echelons. If forced to choose, racer/cruisers are much more likely to buy another spinnaker than have the refrigerator repaired.

Cruiser/Racers

Here the emphasis is somewhat more relaxed. These people are primarily concerned with cruising, but they like to go out and do battle when the sap is rising. A lot of sailors fall into this category, recognising that racing is an excellent way to hone sailing skills. Almost any boat – except perhaps the most lumpen plodders – can be raced on an occasional basis.

Cruisers

It could be said that anyone who doesn't race is automatically a cruising sailor, but I believe we can usefully tighten that definition. Opinions will no doubt vary but, for the purposes of this book, I am regarding cruisers as people who like to travel under sail – sort of water-borne tourists. Beyond a certain point, distances become unimportant; a trip along the coast is as legitimate a cruise as a trans-ocean passage. The characteristic that distinguishes the cruiser is mobility not range – though the urge to go further is often a prominent factor. Cruising sailboats can be of virtually any type, but tend to be conservative, though there is a shift towards more exciting concepts.

Although certainly a large group, cruising people are not as numerous as one might imagine. To own a cruising boat doesn't instantly make one a cruising sailor. It is the way a person sails which qualifies him.

Weekenders

As distinct from the cruising sailor who sets off for a short trip because that may be the only time available to him, the weekender never ventures far from home. His preference is to stay in the marina and potter about his boat, gently sallying forth when conditions are exceptionally propitious. This is a convivial form of boat ownership which should not be disparaged. Often it leads on to more adventurous cruising. But, in the meantime, it places few demands upon the boats, which are chosen mainly for their accommodation and natty appearance.

Live-aboards

Sometimes there is a very fine line between these and cruisers. Obviously, all long-distance yachtsmen qualify as live-aboards because their boats are also necessarily their homes – at least for the duration of their voyages. But the people I have in mind are those for whom the potential mobility of their boats is secondary to the habitat they provide. Many live-aboards regard sailing as painful interludes between one location and another, to be minimised where possible. Progress tends to be leisurely in the extreme, so good performance is rarely important. On the other hand, carrying capacity is always a priority. The typical live-aboard yacht is usually of heavier than average displacement.

Interestingly, this group is a potent source of influence for other sailors – reinforcing the attraction for the type of boats they favour. Many people dream of selling up and setting sail, and the live-aboard – because he always seems to be there – is sometimes the most accessible role-model. The fact that he seldom quite gets around to sailing away himself often escapes attention.

This book is pitched towards those who actually want to go somewhere in their boats, and to do so with minimum aggravation and discomfort. Although much of the information could be of interest to others, it is the sailing yacht as a means of travel that concerns us here. For racing projectiles or floating homes, you may need to read elsewhere.

So, what are these requirements of the travelling sailor? This rather depends upon whom you ask, but for my money they include:

- An efficient hull which is easily driven and well balanced under all points of sail.
- A well designed deck layout giving good protection in the cockpit and secure movement forward.
- A sturdy rig and sensible sail-plan which can be readily managed by a short-handed crew.

- Proper equipment of adequate strength and power for the job.
- A convenient and comfortable interior arrangement, not too ambitious for the size of sailboat.
- A powerful enough auxiliary engine to get you home when the wind dies.
- An electrical supply and distribution system, properly protected against the environment, and of sufficient capacity to handle the demands of modern day accessories and electronics.
- Good navigational facilities and instrumentation.

In approximately that order, we shall cover these and other matters in the following chapters.

There are many different types of rig to be found in different parts of the world, all of which have evolved to suit particular sailing conditions.

Chapter 2
Designed to Sail

Performance

For some reason, good performance – by which, for the moment, we mean speed – is considered indecent by many cruising sailors. Fast boats, they argue, are like fast women – disreputable harlots, worthy only of a faintly curled lip. But a 'proper yacht' (here the tone becomes warm with approval) is a saintly, virtuous lady, of the kind you would happily take home to meet your mother.

This all strikes me as very odd, for there seems to be considerable benefit in completing passages quickly. For example, the difference between averaging five and six knots could shave two hours off a trip from Poole to Cherbourg, and four days off a trade wind trans-Atlantic passage. Expressed another way, you could sail twenty per cent further in any given time – perhaps an extra 250 miles during the course of a three week vacation.

The practical significance of this sort of arithmetic arose for me and my wife, Chele, when recently we were returning from our annual cruise. The final stage was from Lannion, North Britanny, to our home port, Poole – about 140 miles across the Channel along the rhumb line. We intended to do it in a single hop. It was high summer and the days were long. The winds were fresh and favourable and we knew we were in for a fast run. But how fast? Differences in average speed would fundamentally influence the planning of our crossing, as can be seen below:

5 knots: A twenty-eight hour trip, including one full night at sea. Leaving at 0400, to allow us to pick our way through the rocks in gaining light, we would cross the busy shipping lanes north of the Casquets in the dark. With only two people aboard, watches would be essential if we were to remain alert.

6 knots: A little over twenty-three hours. Again, a night at sea, but we would be just about through the traffic before darkness fell. Watches would still be advisable.

7 knots: Twenty hours. Our dawn departure would have us arriving about midnight. The traffic lanes would be crossed entirely in daylight. Full watches could be abandoned. An experienced crew in a well found boat can easily function efficiently over such a period, taking turns to rest on an unscheduled basis.

8 knots: Seventeen and a half hours. Stretching it in our thirty-two foot (9.75m) sloop, but this sort of average could comfortably be achieved by a larger, swifter vessel. A busy day's work for the crew, but hardly an ordeal. You would be alongside well before the pubs closed.

It can be seen that an increase in speed will completely transform the nature of this fairly typical passage. Accomplished at five knots, it would be quite demanding; at eight knots it would be barely more than an exhilarating jaunt. And speed may help you beat any bad weather home. Even in the foulest conditions there are lulls between the blasts, during which a fast boat might safely make a run for it.

If anything, good performance should be of more importance to the cruising sailor than it is to the racer. To be successful in handicap racing, boats have only to perform better than their rating predictions – the last boat over the line can win on handicap. But cruisers live in 'real time', and the implications of their performance is absolute, good or bad. Within reason, speed is a thoroughly good thing.

Boats dance to the rhythms of natural laws, trapped on the interface between two fluids, air and water. There is no opting out from the rules which govern that ever shifting world. All must comply. The trick is to comply well.

Expressed in the simplest terms, a sailing boat is a machine in which the forward propulsive drive generated by the sails must overcome the total resistance of the hull. The more the drive and the less the resistance, the better will be the performance – and, of course, vice-versa. To understand how and why, it is necessary to have some grasp of the limitations involved. It isn't absolutely essential to plough through all of the sums (though, if you have the stomach for it, you would find it useful), but no one can have an informed opinion on sailboat design without taking aboard the basic principles.

We shall be looking at rigs in a later chapter. For the moment let's concentrate on hulls – or, more specifically, on monohulls. The merits of the better multihulls are indisputable but their design is more properly the subject of another book. Single-hulled vessels comprise the vast majority of cruising yachts and it is difficult to see when this would ever be otherwise.

Comparisons

Before we plunge into the technicalities of hull design, it is worth spending some time dwelling on the problems associated with comparing boats of different sizes.

To illustrate this, let's assume that we have a twenty-five foot (7.62m) LWL yacht which we want to scale up to fifty feet (15.24m) LWL, with all other measurements being doubled up similarly. If we did this, would we then have a boat exactly twice the size of the original in every way?

Well, in some respects, yes: Length, beam, and draught would certainly have doubled (times two); but sail area and wetted surface area would have increased by the square (times four), displacement by the cube (times eight), and the stability by the power of four (times sixteen). For we have strayed into that area embraced by the law of mechanical similitude, where not everything is as straightforward as it might seem.

Incidentally, it was this mathematical reality which helped kill off the commercial sailing ship. The economics of carrying freight made bigger and bigger ships desirable. It took almost the same number of men to crew a 500-tonner as one of 1,000 tons. Unfortunately, as displacement increased by the power of three (cubed), sail area was only increasing by the square. Before long, naval architects found that it was impossible to set enough sail to propel the ships they were building. The grotesque six and seven masted schooners were their last desperate attempts to cheat the law of mechanical similitude.

When assessing performance, these matters again arise when we try to compare speeds for boats of different length. For example, six knots would be respectable for a twenty-five foot (7.62m) LWL boat but paltry for one of fifty feet (15.24m) LWL. But a more meaningful comparison can be made by expressing the relative speed of a hull in terms of the Speed/Length Ratio (S/L Ratio). This is determined by dividing any given boat speed (in knots) by the square root of the waterline length (in feet). Metrically, the same results can be obtained by multiplying the square root of the waterline length (this time in metres) by 1.8 before dividing it into the boat speed. Thus:

$$\text{S/L Ratio} = \frac{\text{Speed in knots}}{\sqrt{\text{LWL (feet)}}}$$

or, metrically:

$$\text{S/L Ratio} = \frac{\text{Speed in knots}}{\sqrt{\text{LWL (metres)}} \times 1.8}$$

Now, for reasons which we will shortly discuss, the maximum practical speed for any displacement hull tops out at an S/L Ratio of about 1.4 (seven knots for a twenty-five footer and nearly ten knots for the fifty footer). So, if both boats were making only six knots then the smaller would be at eighty-six per cent stretch whilst the larger would be loping along at just over half its potential. But if the same two boats were sailing at different speeds but identical S/L Ratios, then they would be at the same point within their own speed ranges and would be experiencing much the same hydrodynamic conditions – at least in flat water, ignoring wave size.

The value of the S/L Ratio is that we can now talk about boat speed in purely relative terms without the need to specify lengths. But, from here on, if you want to think in absolute speeds for a specific boat you have in mind, make a note of its $\sqrt{LWL}$ and simply multiply by the S/L Ratios as we go along.

Resistance

The reluctance of a hull to move through the water is caused by the total resistance arising from three distinct sources. These are:

1) Skin drag
2) Wave-making resistance
3) Windage.

For the purposes of this section, we can ignore the windage for now, but let's look at the other two in some depth.

Skin Drag

This is caused by the friction between the underwater surface of the hull (aptly known as its 'wetted surface area') and the water. Frictional resistance is the major drag component at low speeds – as much as sixty-five per cent of the total at an S/L Ratio of 1.0, dropping to about ten per cent at an S/L Ratio of 1.5.

The drag penalty imposed by friction is greatly exacerbated by roughness to the underwater surfaces. It is not the major swoops and hollows that do the damage, but the little imperfections such as weed fouling, protruding skin-fittings, or pitted antifouling paint. The water directly in contact with the hull has some forward velocity imparted to it by friction. This effect is transferred outwards by the interlocking of the water molecules, diminishing to zero at some distance from the hull's surface. This layer of water is carried along with the hull and is known as the boundary layer. Whilst the boundary layer remains attached and laminar – flowing parallel to the hull's surface – frictional drag is minimised; but once it becomes turbulent, drag increases considerably. Laminar flow is easily destroyed by roughness. Even a pimpling of infant barnacles will have a detrimental effect upon light weather performance.

Wave-making Resistance

Nothing is created without cost, and the generation of waves (including all the little eddies and vortices caused by the various appendages that dangle beneath a hull) costs energy. As speed rises, the hull's efforts to shoulder aside the water results in a rapid increase in this form of resistance. Somewhere around S/L Ratio 1.5 it will peak at about ninety per cent of the total. When a displacement type hull reaches this point, it will be sailing at what is known as its hull speed.

The best way to understand hull speed is to know something about the wave dynamics that cause it. The faster that waves travel across deep water, the greater will be the distance between crests. This relationship can be mathematically expressed as:

Wave Speed (knots) = $1.34 \times \sqrt{\text{Wavelength (feet)}}$

or, metrically:

Wave Speed (knots) = $2.43 \times \sqrt{\text{Wavelength (metres)}}$

Imagine a boat accelerating from rest. As it starts to move forward, a bow wave is generated, with a shallow trough and secondary wave forming a little way behind it. As speed increases, so does the wavelength in accordance with the above formula. The secondary wave moves progressively towards the stern. By S/L Ratio 1.34, the secondary wave will have moved right aft and, obviously, the wavelength and the waterline length will then be equal. Now the hull is supported fore and aft by the crests of the two waves. The widest, most buoyant central part of the hull coincides with the trough, and the boat sinks lower. If speed then continues to build, the quarter wave will move clear astern, and the hull will squat still deeper into the trough. From here on it is literally uphill work. Increasingly, between S/L ratios 1.34 and 1.65, the boat will find itself faced ever more with the daunting task of having to sail over its own bow wave – something only the lightest, most powerfully rigged boats can achieve.

This is a compelling argument in favour of generous sail area and light displacement. Although all displacement hulls (when in that mode rather than if planing), will eventually be limited by their hull speed, the exact point where they 'hit the wall' will vary from boat to boat. And clearly there is benefit in delaying this limitation as long as possible. For a boat of twenty-five feet (7.62m) LWL, the difference between S/L Ratios 1.34 and 1.65 is over 1.5 knots – very well worth having!

Planing

Once this was considered the exclusive territory of surf-boards, dinghies, and the lighter multihulls. Monohulled yachts – especially cruisers – were expected to remain respectably embedded in the water, pinned back by the limitations of hull speed. But, it has been known for some time that larger yachts will plane handsomely if designed for the task and given the right conditions. Yet we have continued to discourage it.

Racing is the natural seed-bed for the development of the fastest yachts but, unfortunately, it was in the odious grip of the International Offshore Rule (known as IOR) during much of the sixties, seventies, and eighties. This rule was a hybrid cobbled from the British RORC rule and the American CCA rule, and it exerted a stifling influence on racing yacht design for far too many years. To be advantaged under IOR, it was necessary to design boats with grotesque sectional shapes, pinched sterns, and very little in the way of initial stability – none of which did much to promote genuinely good and seakindly performance. Happily, this is all now astern and the International Measurement System (IMS) and the Channel Handicap System (CHS) have emerged from the wake to shine a much more kindly light on future development. Today's racing yacht is a very fast beast indeed, and designers are naturally delighted that they can turn aside from the sea-lawyer's fine print and can at last concentrate more on speed and efficiency.

Of course, the more extreme racing boats will never make acceptable cruising yachts, but it is inevitable that their influence will slowly change our general perceptions of what is a 'good' sailboat. The modern racing yacht will plane like a dinghy and, as this deplorable behaviour becomes more acceptable, so soon will many cruising yachts. The critical consideration is weight, which takes us on to the next topic.

Displacement

To determine whether a boat is 'light' or 'heavy' for its length, another ratio is employed. This is the Displacement/Length Ratio (D/L Ratio) which is arrived at by the following calculation:

$$\text{D/L Ratio} = \frac{D}{(.01 \times \text{LWL})^3}$$

Where: D = Displacement in tons of 2240 lbs.
         LWL = Waterline length in feet.

D/L Ratios of around 100 would be ultra-light; 200 would still be considered light; 300 would put us into the medium displacement range; and by the time we get to the 450 mark, things are getting seriously heavy. For those who become glassy eyed at the sight of formulæ, I have calculated the various D/L Ratios for yachts of various waterline lengths.

But, however you do the sums, I commend you to heed the D/L Ratio when choosing a boat. The role of weight as a performance killer is readily understood when talking of, say, automobile design or jogging. But amongst sailors there remains an ingrained, almost perverse resistance to accepting what is, after all, an obvious and demonstrable truth. Admittedly, displacement can affect the characteristic way in which each boat rolls and pitches its way through the water, but an excess of it never made a yacht sail fast. The propulsion of mass requires energy – the more mass you have, the more energy you need.

Designed to Sail

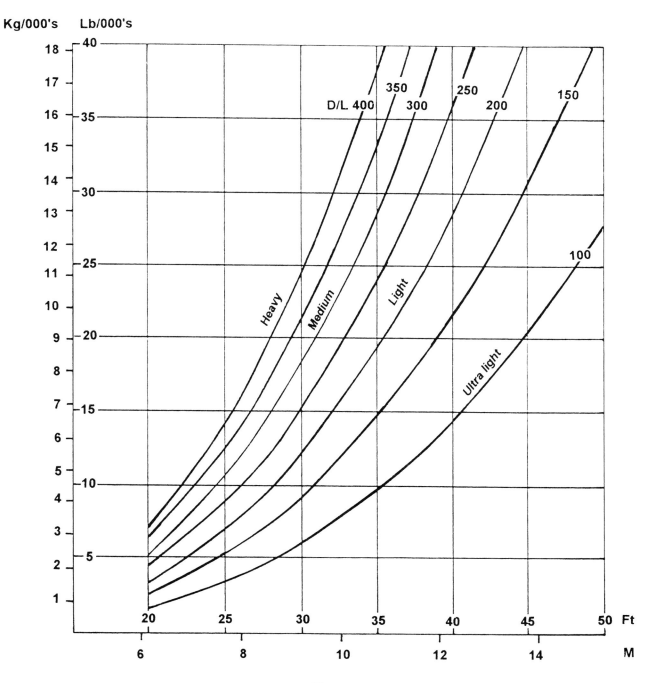

To state the obvious, when not under auxiliary power the motive force which propels any yacht is derived solely from its sails. The larger the sail area, the more power you have at hand. And yet another calculation relates sail area to displacement and gives an indication of how powerfully rigged a yacht might be. In any other field this could be thought of simply as a power/weight ratio, but specific to yachting it is called, as one might guess, the Sail Area/Displacement Ratio. This is derived thus:

$$\text{SA/D Ratio} = \frac{SA}{(D/64)^{0.667}}$$

Where: SA = Sail area in feet.
D = Displacement in pounds.

Most boats will be found to have an SA/D Ratio of between 14 and 20, with the higher figure at the more powerfully canvassed end of this range. Again, for those not blessed with a slide-rule mentality, there is a graph opposite which allows quick approximation for boats of varying sizes.

I would ask you to believe that all of the foregoing is not an attempt to 'blind with science'. If our *machine à naviguer* is to perform creditably in the circumstances for which it is intended, then these and other matters must be optimised by the designer, and should be weighed up with some discernment by any prospective purchaser. No one would buy a car without carefully assessing such considerations as performance and fuel consumption, but it is astonishing how some folk blunder into boat ownership with hardly a notion of what they are getting.

Prismatic Coefficient

Before we close the chapter on performance related design considerations, let us take a look at this vital but obscure little number, the meaning and significance of which is not always understood.

The word 'prismatic' is confusing. The prism in question is actually the maximum underwater section extended longitudinally to equal LWL. The volume of the prism so produced is said to have a value of one (1) and the actual volume of the tapered hull is then expressed as a proportion of it. For instance, a hull whose prismatic coefficient is .50 has fifty per cent of the volume of the prism, a prismatic coefficient of .55 would represent fifty-five per cent of the prism volume, and so on. In practical terms, the prismatic coefficient is a measure of how fine or full are the ends of a hull.

Much of the credit for recognising the importance of prismatic coefficients is due to an American, Admiral David W. Taylor, who was working on the design of warships at the time. He found that for every S/L Ratio there was an optimum prismatic coefficient. For example, a hull travelling through the water at an S/L Ratio of 1.0 would have the least wave making resistance if it had a coefficient of .52. If

Designed to Sail

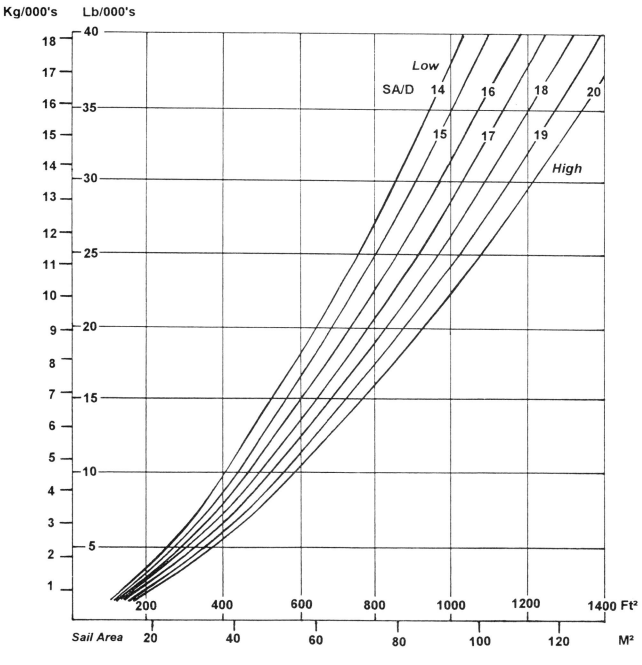

The Innovative Yacht

the S/L Ratio was around 1.4 (hull speed) the ideal coefficient would then be .64. And to further illustrate the importance of the coefficient, consider this: If that last boat with its tubby coefficient of .64 was slopping along at the lesser S/L Ratio of 1.0, its resistance would be approximately double that of the finer hull.

The problem for the yacht designer is that his creation will eventually be sailing at S/L Ratios of anywhere between barely moving and flat out. Somehow he has to decide which speed range his boat will most often be performing in and choose the prismatic coefficient accordingly. If he knows the boat will sail in an area where light airs predominate and speeds will generally be low, he may choose a lower than average coefficient – and, of course, vice-versa. To some extent this accounts for why yachts can gain reputations for being 'dogs in light airs' or 'demons when it blows', and also why a yacht can be startlingly successful in one area and a disappointment in another.

The inescapable compromise a yacht designer makes when he selects the prismatic coefficient is part of his 'black art' – the combination of technical knowledge and empirical guess-work. And of all the design details he is likely to release, this is amongst the last he will want published.

Therefore, to summarise: the factors that most profoundly influence performance are:

Waterline Length
The longer the better, though this could incur greater wetted surface area which can be deleterious in light conditions.

Wetted Surface
The less the better, particularly in regions where light conditions predominate.

Displacement
Within reason, also the less the better, though load carrying and structural requirements must obviously prevail.

Prismatic Coefficient
Optimised for conditions and usage. Most cruising designs will have coefficients compromised to provide the broadest possible range of operational efficiency.

Sail Area
The more the better, again within reason. Remember, you can always reduce sail as the wind rises. To have to reef early is nothing like as exasperating as wishing you had an extra metre on the mast in light airs.

Handling

But performance isn't all about speed. Nothing establishes a cruising boat so firmly in the affections of its crew as the way it handles. The well tempered yacht will have praise heaped upon it; the cantankerous, cranky, hard-mouthed sea-cow (of which, alas, I have known too many) will be bitterly reviled. Good handling characteristics are especially vital for the short-handed crew. The racing helmsman may relish fighting the tiller as he blasts around the cans, but the sensation of having arms pulled from sockets is not something to be savoured for more than a few hours.

The blend of interacting factors that combine to make up a seakindly yacht call for the subtlest skills of the yacht designers' art. To design a fast boat is one thing – to design a fast and mannerly boat is quite another. And here we are referring not only to sailing, but manoeuvrability under power as well. Watch the fleet returning to any marina at the end of the day. You could see more entertainment and embarrassment compressed into a shorter time span than anything you might see offshore.

The cruising yachtsman should demand three principal handling qualities:

- A yacht which is easy on the helm under the widest possible range of conditions; one which will not gripe up unduly as the angle of heel increases. Of course, achieving the right balance is largely due to sail trimming, but some boats defy all attempts to tame them. Although most behave reasonably well in light conditions, the distinction between swans and geese will assuredly emerge when the wind pipes up.

- Good directional stability both on and off the wind. To be able to leave the helm, even for a few seconds, is a great boon. An important bonus is that a directionally stable boat will make light work for any vane gear or autopilot.

- Nimbleness when manoeuvring under sail or power. Ironically, this is the antithesis of directional stability – a classic case of wanting it both ways. But the 'point and pray' type of yacht, which takes a heart-stopping age to respond to the helm, can be wonderfully docile offshore but a nightmare in confined circumstances such as marinas.

Rather than analyse what each feature of a yacht's design might contribute to individual handling traits, let's take a look at three clearly distinctive types and assess how we could expect them to perform and handle.

Heavy Displacement

The first example (Fig. D1) is an older type of yacht with a long keel and transom hung rudder. Obviously, such a boat will certainly be of heavier than average displacement (D/L Ratio from about 350 upwards). Her high wetted surface area will make her slow in light airs, and her massive weight will probably pin her S/L Ratio to no more than 1.4 at a push. Although capable of carrying a powerful rig, this is often not the case, perhaps because the limitations of her hull form rarely warrant it.

With her long keel she could be expected to 'run on rails' – about as good as you can get in terms of directional stability. However, her manoeuvrability will be extremely ponderous, and she will be almost uncontrollable when going astern under power.

The Innovative Yacht

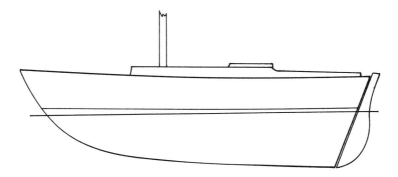

Fig D1 Heavy displacement yacht

When heeled, the general symmetry of her inclined waterlines should make her well balanced, though the large area of her unbalanced rudder will make her heavy on the helm at all times, whether heeled or not.

Her motion through the water should be as comfortable as it is sedate. The markedly veed forward sections will give a very soft ride, though she will tend to bury her bow and could be extremely wet.

This type of yacht is a good load carrier but a poor performer. Perhaps ideal for live-aboards and those who admire tradition, but rather inefficient as a cruising boat for sailors genuinely on the move.

Medium Displacement

(Fig. D2) With a D/L Ratio of around 300, this group once occupied the vast middle ground of cruising yachts, but is now beginning to look a little heavy by modern standards. The typical medium displacement boat will have a moderate length keel, and either a skeg or transom hung rudder. Her overall beam is likely to be modest.

Performance cannot be expected to be scintillating, but should be fair enough in most conditions. Manoeuvrability, directional stability, and balance will be dependant upon the quality of the design; there is no reason why all should not be excellent, but there is broad variation between classes. Good medium displacement yachts are often the most tractable boats to handle.

Because living space and load carrying ability are both volumetrically related, medium displacement is a good choice for smaller yachts, but becomes less attractive as size increases and an abundance of volume becomes available (by the cube, remember).

Designed to Sail

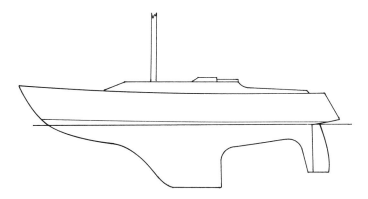

Fig D2 Medium displacement yacht

Light Displacement

(Fig.D3) More and more yachts fall into this category, typically with a D/L Ratio hovering around the 200 mark. This is partially due to economics – less materials means less cost – but also to the awakening public interest in better performance.

The light displacement yacht will have an uncomplicated hull shape rather like a large dinghy, with a long waterline and fairly low wetted surface area for her length. Her ballast keel will probably be a medium aspect ratio fin or – in the UK, where uniquely they remain popular – twin bilge keels. The rudder might be hung on the transom, on a small skeg, or be of the cantilevered spade type.

As light displacement obviously denies the carrying of a lot of ballast, stability is usually gained by increasing the beam. If excessively heeled, the resulting asymmetry of the waterlines associated with broad aft sections can make them hard on the helm (but, again, this is very much down to the skill of the designer). Performance should be brisk in nearly all conditions – especially off the wind, when hull speed may well be exceeded.

Light boats are relatively disadvantaged when beating to windward in heavy going. Lacking the weight of heavier vessels, they are not as authoritative when punching through the waves. But the notion that they cannot be driven upwind is nonsense; they can, but somewhat less comfortably. Flatter forward sections tend to pound, and the ride can be on the lively side.

Under sail, they are invariably easy to handle, tacking precisely and without hesitation. Manoeuvrability under power – both ahead and astern – should also be very good; most will spin on a sixpence. However, when motoring at low speeds, they can be skittish in strong cross-winds. This type of yacht requires emphatic handling. Maintaining steerage way becomes especially important when coming alongside. Too tentative an approach can see you blown off your line, possibly with amusement to any bystanders but with ignominy to yourself.

The Innovative Yacht

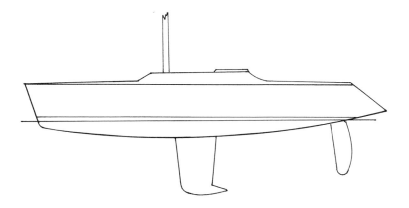

Fig D3 Light displacement yacht

Clearly, if you overload any light displacement yacht, you will erode the performance advantage it offers. And the smaller the boat, the more acutely this will be felt. To illustrate this, let's imagine two boats, both of D/L Ratio 200, but of different lengths – one at twenty-five feet (7.62m) LWL and the other at forty feet (12.19m) LWL. Fuel, water, and provisions for a long voyage are put aboard, adding 1,500 lbs (680.4 kgs) to the displacement of each. This will lift the D/L Ratio of the larger boat up to 210 and the smaller to 242 – an obviously more burdensome penalty.

Ultra-light Displacement Boats (ULDBs)

These have many of the same characteristics as the previous group but more so. As you would expect, performance can be spectacular. Under the right circumstances, ULDBs will plane like dinghies, easily tossing aside the limitations of hull speed.

But there is a price to be paid. Very light weight may call for exotic building techniques, which rarely come cheaply. And these boats are inclined to be fragile, so longevity (and thus the protection of your investment) cannot be guaranteed.

Although the larger ULDBs are finding favour amongst some cruising sailors, their use is more usually limited to racing – often of the long distance, epic variety. As such, these extreme monohulls are rather outside the scope of this book.

Insofar as sailing characteristics are concerned, the question of displacement – or, more correctly D/L Ratio – is perhaps the most crucial the prospective buyer should consider. Nothing affects the 'feel' of the boat so much, or will determine how it will behave in various conditions. Naturally, the final choice will be up to you – and it isn't an easy decision. Whilst heavier boats will have a more comfortable motion, they will turn in slower passage times and be awkward to handle under power. ULDBs may be blisteringly fast but you may need a resident chiropractor as part of the crew. Somewhere between them is the perfect compromise for you.

Designed to Sail

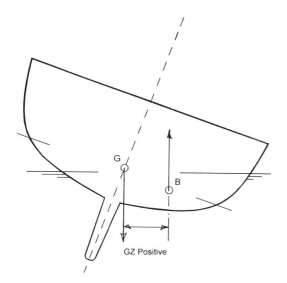

Fig D4.1 Stability —stable

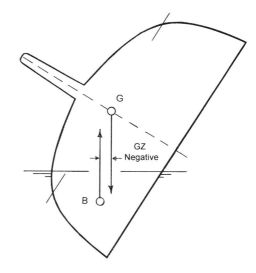

Fig D4.2 Stability — unstable

Stability

The stability of a ballasted single hulled vessel is derived from two sources: the sectional shape of the hull (principally, beam), which is known as its form stability, and the weight and position of the ballast. Some boats – multihulls and most power craft, for instance – carry no added ballast, and must therefore rely entirely upon form stability. But for now we are concerned with the conventional sailing yacht.

The purpose of ballast is, of course, to lower the Centre of Gravity (G) of the entire structure. In dynamic terms a moment is involved – the efficacy of the ballast being a product of its weight and its vertical distance beneath the Centre of Buoyancy (B). The heavier the ballast or the lower it is placed, the more righting effect it will impart.

The effect of beam is more obvious: expressed simplistically, the wider an object is the more difficult it is to push over.

Figure D4.1 shows the way the various forces work. A typical modern hull section is shown heeled at 20°. B (which, at rest would be on the centreline) has moved to leeward and is exerting an upward force. G is now to windward of B and is pulling downwards. The distance between the vertical lines through which these two forces are said to act is another moment commonly known as the Righting Arm – or more technically as GZ.

As the boat heels more, GZ will increase until it eventually peaks at between 60° and 80°, depending upon the type of boat. Beyond this point of maximum stability GZ will decline until it becomes negative, whereupon the boat will capsize (Fig.D4.2).

The Innovative Yacht

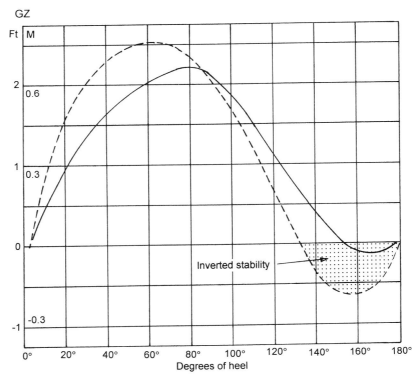

Fig D5 GZ curve

Yacht designers calculate GZ for various angles of heel and plot the results on a graph. This produces a GZ Curve, which is an excellent visual indicator of the stability characteristics of any design.

Two typical GZ Curves are shown in Figure D5. These are worth studying, as they illustrate crucial difference between two common types. The dashed line is a representative modern cruising yacht with a fin keel and fairly wide beam. The solid line is a more traditional yacht of narrower beam and heavier displacement. It can be seen that the modern boat stiffens up early and reaches its maximum GZ at about 60°. The traditional yacht, on the other hand, is initially more tender and doesn't achieve its maximum stability until nearly laid on its ear (the curve, incidentally, can be seen to flatten slightly when the rail is driven under, to steepen again as the coachroof contributes to buoyancy).

Of chilling importance to offshore sailors is that part of each curve which lies beneath the Zero GZ line. This represents the inverted stability of the boat; the more curve there is on the negative side, the less eager it will be to recover from capsize. One of the main criticisms levelled at multihulls is that they are as stable upside down as the right way up – more so, indeed, if the rig remains intact. What is not always appreciated is that some monohulls – especially those burdened with radars, furling gears, and other gadgetry aloft – could also be fairly happy bobbing around with their keels in the air.

Comparing our two boats, it can be seen that the modern boat certainly has some inverted stability (though would still probably right itself fairly quickly in a rough sea) whilst the traditional yacht would bound to her feet almost without hesitation. Remember, these curves represent static stability; the dynamic case, involving wind and wave action, is quite another matter.

The designer is confronted with a selection of trade-offs which he must blend to suit the purpose of the boat. To gain stability he can either:

- Opt for a high Ballast Ratio (which is the weight of ballast expressed as a percentage of that of the whole boat). This he may consider undesirable if he is shooting for light displacement.

- Place the ballast as low as possible, either by deepening the keel or concentrating the ballast in a bulb. Here, for a given righting moment, weight would be saved, but any increased draught may be unacceptable. Some structural problems might also arise.

- Choose a beamier hull shape. This could make the yacht a bit of a tub, hard on the helm when heeled and uncomfortable to windward in lumpy seas. Also, in extreme cases, the inverted stability might be a worry.

As they say in the north of England, 'you don't get owt for nowt'. However intelligently you interpret the triangular relationship between weight, draft and beam, gains in one direction will demand sacrifices in another.

Water Ballast

From the traditionalists, I can hear the howls of protest already. But, for some applications, I believe this is worth a very serious look – especially if you intend medium to long distance cruising, where short tacking is rarely the norm. To be able to supplement the permanent ballast with additional weight to windward has powerful appeal. The racing boys, with their legs dangling outboard, have known this for a long time.

Figure D6 shows a schematic of the arrangement aboard one of my designs – the RQC38. Wing tanks are flooded and drained with sea-water by a combination of natural gravity and pumps. The flow is controlled by simple valves. At 10° heel, 1,000 lbs (450kg) of water ballast to windward would be worth about an extra 3,300 lbs (1,500kg) of lead on the bottom of the keel! The benefit lessens as heeling advances, but it still remains very useful over normal sailing angles. Of course, there must

The Innovative Yacht

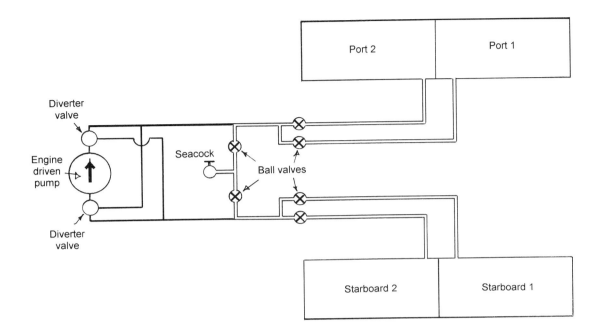

Fig D6 Water ballast

always be sufficient permanent ballast to provide reserve stability without any water ballast or – the worst possible case – with a full ballast tank to leeward.

The attraction of such a scheme is that it is totally adaptable to each set of circumstances. You can carry no water ballast at all in light weather or when short tacking. But once set on a longish leg, you can progressively pump water to windward to help keep the boat on its feet. Apart from making her more stable, this also increases the D/L Ratio, providing the extra punch to take her through the waves.

Although there is a price to be paid in extra work, this is no more demanding than, say, a sail change. It is likely that water ballast will appeal mainly to more experienced sailors, who will relish the opportunity of being more in control.

Insofar as sailing characteristics are concerned, the question of displacement – or, more correctly D/L Ratio – is perhaps the most crucial the prospective buyer should consider. Nothing affects the 'feel' of the boat so much, or will determine how it will behave in various conditions. Naturally, the final choice will be up to you – and it isn't an easy decision.

Chapter 3
Aloft

The American journalist and cynic, H.L.Mencken, once said of democracy that it was 'based on the notion that the people knew what they wanted – and deserved to get it good and hard'. The same could be said of rigs, where sentiment and fashion too often prevail over good sense.

A slippery hull deserves an efficient rig. Again, there is a tendency for cruising sailors to settle for less, but this is as illogical as the 'cruisers must be plodders' argument. As we have previously discussed, speed has its own benefits; and the power to produce this is drawn from the sails.

Before we examine the different types of rig, let's take a look at the sails themselves – what they are and how they are put together.

Sails

Sails are aerofoils which convert one form of motion into another by generating lift. The more effectively they do this, the greater the benefit for any sailor – the cruiser as much as the racer. Figure D7 shows three types of rigid foil presented at the same angle of incidence to the wind: (a) is a flat plate, (b) is the same plate, but now slightly cambered as a sail might be, and (c) is a thicker asymmetric foil, similar to an aircraft's wing. The ratio between lift and drag (no chance of anything for nothing here either) is noted below each type and can be seen to improve dramatically between the crude (a) and the refined (c).

It's fairly easy from this to imagine such a foil set upon a hull, and trimmed as if hard on the wind. Lift and drag resolve to produce both the Forward Drive Component and the Heeling Component, as illustrated in Figure D8. Although lift contributes to both drive and heeling, its effect obviously has overall benefit and should be maximised. Drag on the other hand has a negative effect when beating and, consequently, the less of it the better. However, as the apparent wind moves aft of the beam, drag

The Innovative Yacht

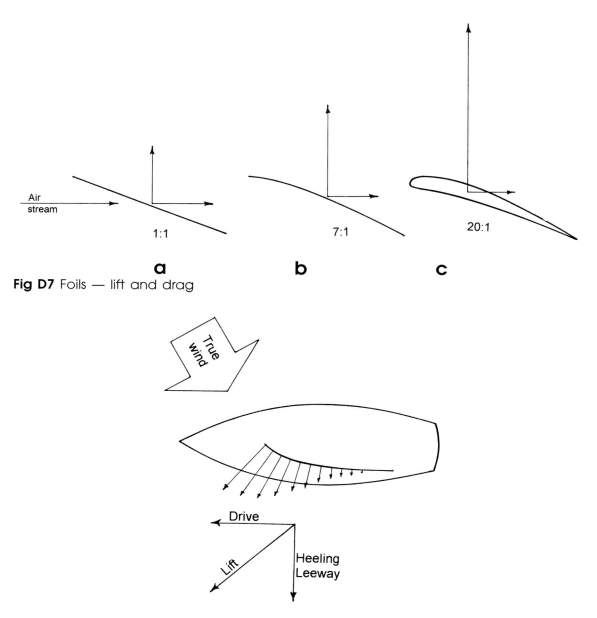

Fig D7 Foils — lift and drag

Fig D8 Foils — lift and drag, forward drive and heeling

ceases to be a liability and starts to contribute to drive – one of the reasons why a beam reach is the fastest point of sailing. When sailing dead downwind, the function of the sails is to simply stop the wind, and both lift and drag are entirely beneficial.

Reference to our three foil types (a, b, & c) shows the potential there is for improving the Lift/Drag Ratio by varying the shape and trim of our sails. Attempts have been made to introduce rigid foils (c) into sailing, but these have so far failed to live up to the often vaunted claims – largely because of weight, mechanical complexity, and the fact that they can only be optimised for a narrow band of conditions. Most of us will settle for something more like (b) – a soft sail where the critical adjustments of camber, twist, and sheeting angles are controlled by nothing more complicated than bits of string.

The great enemy of sailmakers is stretch. And the degree by which woven sailcloth stretches varies according to the direction in which load is applied. If aligned with the threads – along the longitudinal 'warp', or transverse 'weft' (often called 'fill' in the US) of each panel (Fig. D9.1) – then stretch is minimal; but if loaded diagonally across the 'bias', the cloth will yield and deform (Fig.D9.2). Not all cloths have a balanced warp/weft mix. Some have their strength warp oriented and others weft oriented, depending upon the number and size of threads running in either direction. The construction of the weave also affects orientation. For example, if the weft threads are allowed to lie straight, with the warp threads crimped around them (Fig.D10), a weft oriented cloth is produced. The stability of sail-cloths is usually also improved by adding resin fillers, which literally bind the threads together, but this is only ever partially successful and, if overdone, makes for a very stiff and unmanageable sail.

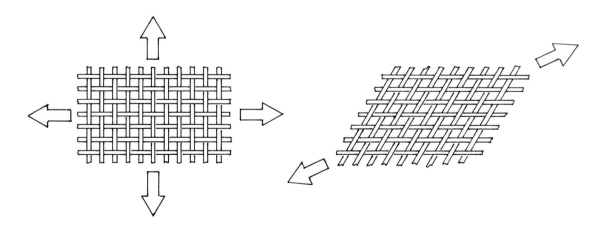

Fig D9.1 Warp and Weft   **Fig D9.2**

The Innovative Yacht

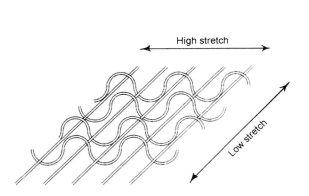

Fig D10 Warp and weft — crimp

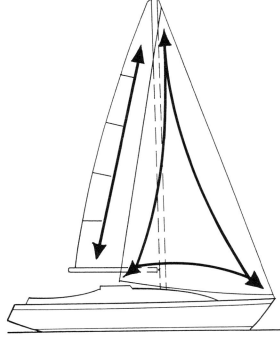

Fig D11 Sail stresses

Sail designers therefore contrive to arrange the cloth panels so that either the warps or wefts (depending upon orientation) coincide as far as possible with the stress pattern for each type of sail. Figure D11 shows a typical sloop rig on which is superimposed an approximation of the direction in which the principal stresses run. To accommodate these stresses, cruising mainsails are usually 'cross cut' with weft oriented cloth (Fig. D12), but 'radially cut' (Fig. D13) mainsails are also becoming common.

Headsails also can either be cross cut or radially cut (Fig. D14.1&2); this choice being strongly influenced by the size of the sail. Radially cut sails hold their shape better than cross cut, and this improved dimensional stability becomes increasingly meaningful as size increases. But radially cut sails are extravagant in both labour costs and material wastage, and many sailors feel that on boats of less than about thirty-five feet (10.67m) the marginal performance gains are hardly worth the extra money. The exception to this would be spinnakers and other light downwind sails, with which we shall deal in a separate section.

Developments in racing sails have left their mark on cruising yacht design, and this downward fertilisation continues on an ongoing basis. Ultra low-stretch materials such as Kevlar™ and Mylar™ are commonplace on racing boats and can now

Aloft

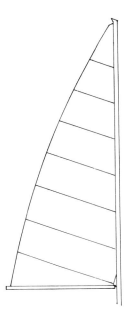

Fig D12 Cross cut mainsail

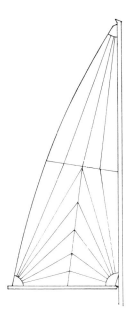

Fig D13 Radially cut mainsail

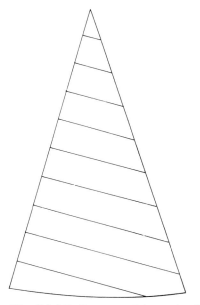

Fig D14.1 Cross cut headsail

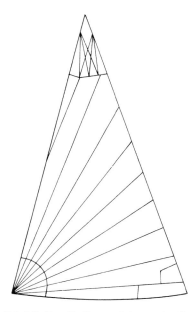

Fig D14.2 Radially cut headsail

sometimes be seen aboard cruisers. Another area of progress centres on laminated sails – a bonded sandwich of Mylar film and polyester cloth, offering superior stability and good wear resistance. Such techniques could make significant inroads into the cruising market over the next few years – most notably in combination with reefing gears, where sails are stowed rolled rather than stuffed into bags.

But, for now, most cruising sails continue to be made in ordinary woven polyester sailcloth. As a material it is strong and reliable; but it is susceptible to attack by ultra-violet light and, in tropical regions where the sun is harsh, this can be very severe. Although not entirely effective, virtually transparent (they give the cloth a slightly creamy tint) coatings have been developed to screen the underlying sailcloth from UV light, thereby usefully prolonging its life. But be warned: there's no point in paying a premium for UV resistant cloth if your sailmaker doesn't also use UV resistant thread! Incredibly, some don't. Check it out before ordering.

Rigs

Except on a few eccentric designs, sails are set not singly but in combinations. Each sail acts individually as an aerofoil and also in concert with its fellows. The entire sail plan, viewed as a whole, can be considered to have a collective characteristic so far as lift, drag, heeling and drive are concerned.

The well designed modern rig is a marvel of efficiency and flexibility. But, human nature being what it is, we continue to strive to re-invent the wheel. Scores of dotty ideas have sprung up over the years, claiming all manner of advantages; but fabric sails, controlled by ropes and hung on aluminium spars, still remain the firm favourite.

The ideal cruising rig should provide:

- Good aerodynamic performance on all points of sailing. Clearly, the more efficient a rig is, the less sail area (and cost, weight and windage) you will need to drive the boat forward. Most rigs perform acceptably off the wind; it is on the beat to windward where the real gains are to be enjoyed.

- Adaptability for different wind strengths, either by reefing or changing sails. Reliability, convenience of operation and crew security are important considerations.

- Ease and precision of control. There's no point in having a superbly cut sail if you cannot trim it properly. On a centre cockpit cutter I recently sailed, the correct sheeting position for the staysail was a point in space inboard of the coaming. Once we had rigged a barber hauler, the sail set splendidly but it became difficult to move about the cockpit without first training as a limbo dancer.

Aloft

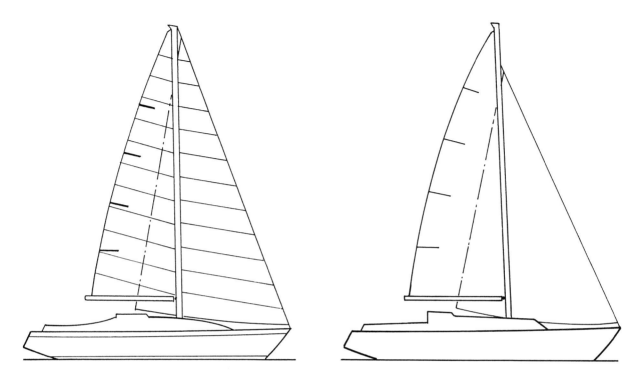

Fig D15 Mast-head sloop   **Fig D16** Fractional sloop

The virtues and vices of any rig are determined by both its general and detailed design. Before moving on to the nuts and bolts, let's compare the most common cruising rigs:

Sloop

The archetypal yacht rig, having a single mast which carries a mainsail and a headsail. This is a deservedly popular arrangement, offering simplicity and efficiency at the least practicable cost.

Sloops come in two variants: 'Mast-head' (Fig. D15), the most common, where the forestay goes to the top of the mast, and 'fractional' (Fig. D16) where it attaches at some distance beneath it (typically, ⅞ of the distance up from the heel). Aerodynamically clean, both types are powerful performers to windward.

The simple features of the masthead rig are well understood, but the less obvious advantages (as some would claim) offered by the fractional rig deserve mention. These are:

- The mainsail is the easiest sail to control and forms a greater proportion of the total sail area on this type of rig.

- Headsails and spinnakers are smaller. Weaker crews can handle them with lighter gear. And when stowed below they take up less space.

- Tensioning the shorter forestay is easier, improving headsail set and the efficiency of roller reefing gears.

- When not under spinnaker or cruising chute, the fractional rig is comparatively better downwind. The big mainsail provides most of the motive power, and it is less important if the headsail is blanketed.

- For the gung-ho boys, a fractional rig allows mast bend to be altered, thereby controlling the fullness of the main. Although very much a racing technique, to flatten the mainsail in heavy weather can be useful in cruising, delaying the point where you must reef.

On the debit side, a fractional rig usually requires swept spreaders, which are more tricky to tune, and running backstays which are considered just too much hassle by many cruising sailors.

Although the robust simplicity of the masthead sloop will probably make this the first choice for most sailors, the fractional rig deserves serious consideration for its own special benefits – and could actually prove the preferred option for some purposes.

Cutter

Another rig with a single mast (Fig. D17), the cutter has always been popular amongst long-distance sailors, and is today enjoying something of a revival in less epic circles. It can be slightly less efficient to windward than the sloop, but really comes into its own when reaching or running.

The distinguishing feature of the cutter is the staysail carried on an inner forestay. This division of the total foretriangle area into two makes the individual sails smaller and easier to handle, but incurs some added complexity. Obviously, more gear is required to deal with two headsails, and both must be re-trimmed every time when going about – though the staysail can be made to self-tack with the use of a boom or athwartship track (Fig. D18).

A hybrid arrangement (which once, but thankfully no more, laboured under the charmless name of 'slutter') can be either a sloop or cutter as required. Here the inner forestay can be detached and moved aside to clear the foredeck for tacking large genoas, and re-rigged when off the wind in heavier conditions.

The glory of the cutter lies in its versatility. Figures 19.1,2,3 & 4 show how progressive reductions in sail area, from light to storm conditions, might work in a typical arrangement. It can be seen that not only does the sail area shrink at each

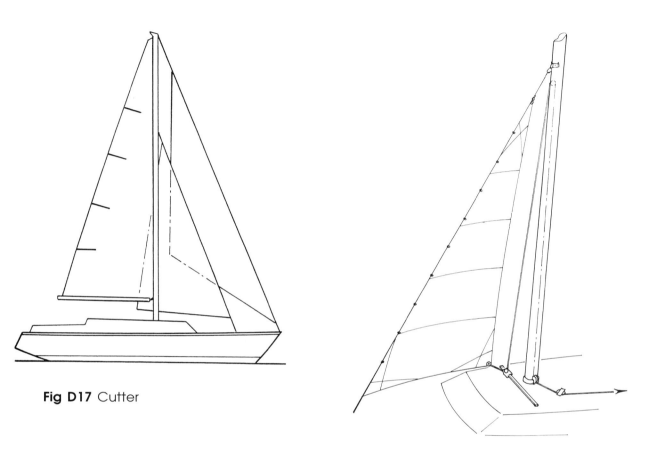

Fig D17 Cutter

Fig D18 Self-tacking staysail

stage, but it also moves inboard where it can be more safely handled. This contrasts with the sloop where the storm jib would remain hanked-on right forward on the exposed foredeck. Another advantage is that the mast is extraordinarily well supported. The deep reefed main (or storm trysail) is within the running backstays and can be tacked without first releasing the leeward runner. The security of this rig is wonderfully reassuring for the voyager.

Although probably not worth all the extra trouble in very small yachts, or if you intend to sail mainly in confined waters, the cutter is a very serious contender for cruising offshore in boats upwards of, say, thirty-two feet (9.75m) LOA.

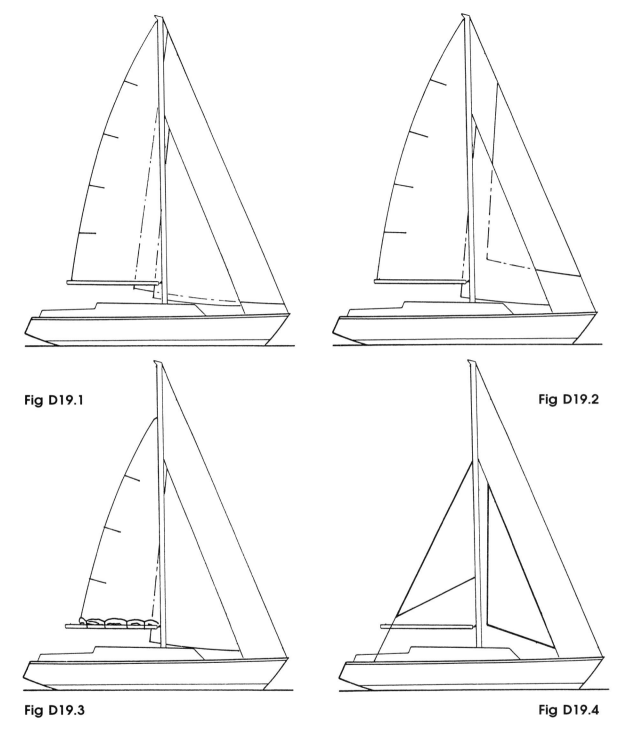

Fig D19.1

Fig D19.2

Fig D19.3

Fig D19.4

Fig D19 Progressive reduction of sail area on a cutter

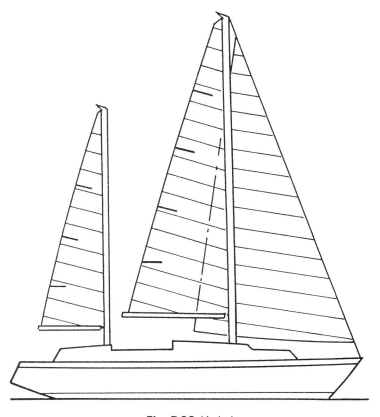

Fig D20 Ketch

Ketch

A two-masted rig, beloved by those who fancy its big boat appearance, ketches are often a romantic rather than a rational choice – though they make much more sense as boat size increases. Figure D20 shows a typical sail plan.

Ketches are relatively poor performers to windward. When beating, the mizzen can be so severely back-winded by the mainsail as to be virtually useless. On this point of sailing they can almost be looked upon as under-canvassed sloops. But once the wind frees, the ketch starts to look better. In light weather, mizzen staysails (or even mizzen spinnakers) can be set, usefully adding to sail area. The effectiveness of ketches in predominantly off-wind sailing was vividly demonstrated in the 1989/90 Whitbread Round the World Race, when the two maxi ketches *Steinlager* and *Fischer and Paykel* led the rest of the fleet home.

Another advantage is that the fore-and-aft spread of the rig contributes to good balance, with plenty of scope for adjustment to keep helm loads light. And at anchor with just the mizzen set and sheeted amidships, the ketch will lie as serenely as a slumbering duck.

Supporters of the ketch will often defend their choice by pointing out that the rig is less strenuous to handle because the sail plan is split into smaller, lighter units. But, whereas this might be significant on big boats, it is hardly a factor for your typical family cruiser where even the largest sail is no problem for a reasonably fit crew.

The counter argument centres not just on the ketch's dubious windward ability, but also on the weight and windage aloft and the extra pennies required to put it there. Personally, I believe that unless there are compelling reasons to opt for the contrary, most smaller modern ketches would have made better sloops or cutters.

And what else?: Not a lot is new – in principle at least. The Chinese lugsail (junk) and lateen rigs have been around for thousands of years, the schooner, yawl, and sprit-sail barge for centuries. All rigs have their devotees, otherwise they would have been long forgotten. And it would be a rash man who would be too dismissive about concepts that have served us so faithfully.

But for the recreational sailor, the modern type of sail plan is very hard to beat. Extruded aluminium spars, stainless-steel standing rigging, and ropes of synthetic fibre might not twang the sentimental heart strings quite so potently as varnished spruce and tarred hemp, but viewed in purely practical terms they are miracles of convenience.

Progress from here will probably be incremental, drawing step by step on advancing technology. Perhaps the sailboard is the only truly new sailing innovation we have seen in the last half of the Twentieth Century. With its articulated mast and wishbone boom, and its amazing method of steering by varying the sail balance, it is stunning in its conceptual freshness.

Unstayed Masts

Although these have been around for thousands of years – most notably on the Chinese junk rig – it is only fairly recently that modern materials have made these an attractive proposition for the yachtsman. In engineering terms an unstayed mast is a simple cantilever which must withstand all the loads imposed on it by the strength of its section alone. Built in timber or aluminium the weight aloft can be prohibitive – often mitigated in many designs by making the mast heights short, thus producing stumpy, underpowered sail plans.

With the arrival of GRP, it was not long before attempts were made to produce laminated hollow spars from this remarkable new boatbuilding material. At first these were only partially successful. The structural qualities of glass fibre and polyester resin are not ideally suited to this application and failures were frequent. Then came epoxy resins, which opened the door still further, but it was not until such high-strength reinforcements as carbon and aramid (Kevlar) fibres became available that the full potential began to be realised. Today, most composite spar makers choose carbon fibre and epoxy, with which they fashion masts that combine immense strength with acceptably light weight. This has opened up a fascinating area of development which will increasingly impact the sailing scene as more designers first recognise, and then explore, this intriguing approach to cruising yacht rigs. Some believe that within a decade or so the conventional stayed rig, with all its struts and wires and other bits and pieces, will seem as outmoded as the early biplanes.

One advantage of unstayed masts is that they can be made very aerodynamically clean, with no rigging, spreaders, or other fittings generating unwanted turbulence. Another is that the punishing compressive loads of conventional masts are entirely absent – though the mast step and deck aperture must obviously be made very substantial to form a secure 'socket' for your cantilevered stick. Less attractively, the inherent flexibility makes proper headsail tension very difficult to achieve. This limits all but downwind headsails to relatively small sizes or, as is often the case, the rigs use no headsails at all.

I think we shall hear a lot about unstayed masts in the future. Simplicity is always a desirable goal of design and there is an obvious visual and structural elegance about having something which will stand up on its own. Of course we still have much to learn, and the stayed rig has by no means outlived its usefulness, but my guess is that, for cruising yachts at least, the days of standing rigging just may be numbered.

Reefing

The ability to match sail area to wind strength is absolutely vital for the offshore yacht. Wind pressure varies with the square of velocity. This means that if wind speed were, say, to double ($\times$ 2), the pressure would increase by $2^2 = 4$.

To underline this point, let's take a modest family yacht with a sail area of 500 sq.ft (46.45 sq.m) slipping along in a gentle Force 3 breeze: the total pressure of the wind in the sails would amount to some 200 lbs (91 kgs). Now, along comes a squall and, before the crew can ease the sheets, the wind rises suddenly to a lusty Force 8 – four times the velocity it was a moment ago. The wind pressure would leap by 4^2 = sixteen times, or a total of 3,200 lbs (1,451 kgs) in the whole rig. It's very easy to see how such a boat could be knocked flat.

Incidentally, although we are talking about sails here, it should be remembered that these escalating effects are felt by every bit of the vessel above the waterline – topsides, coachroof, rigging, the lot. And, while we can reduce sail when the wind gets up, there is nothing we can do about the vessel's permanent windage, except keep it to the minimum at the design stage. A boat with a lot of top hamper may soon find that the drive from its reefed rig is not enough to overcome the adverse wind pressure on its structure. It is largely this phenomena which has blighted the windward reputation of bulkier multihulls.

But, back to sails. In response to demand, reefing systems have seen vigorous development over the past couple of decades, with the emergence of some exciting new gear to make life easier for the sailor – although 'new' is not strictly correct, because most have evolved from older concepts. Some of these systems employ fairly complicated (and costly) mechanical devices; others are hardly more than alternative methods of rigging the various control lines.

Genoa Roller Reefing

A neat and convenient method of reefing and stowing a headsail, which has earned widespread approval amongst cruising sailors. An extruded aluminium spar carries the sail and is rotated around the forestay, driven either by a drum and line system on small to medium sized yachts, or by hydraulic or electric motors on larger vessels where muscle power would be insufficient.

It has to be admitted that the set of a partially furled genoa is not as good as might be achieved by changing down to a smaller jib. As the sail rolls up around the spar, the tension along the luff is lost, and the sail starts to bag – just the opposite of what you want when the wind gets stronger. In order to help overcome this problem, some of the more expensive gears utilise a double swivel arrangement just above the furling drum, which takes a complete roll out of the middle portion of the luff before the head and tack begin to be rolled in. This flattens the sail as reefing commences, though the advantage gradually diminishes as the reef becomes deeper.

Meanwhile, the sailmakers had not been idly standing by. Roller reefing gears create their own demands and the sails must be specially cut to suit them. Converting a conventionally hanked sail to roller reefing is rarely successful.

A foam insert sewn into a pocket at the luff is one of the tricks sailmakers have developed. This fills out the bagginess as the sail is rolled up, thus helping to maintain luff tension as long as possible. In radially cut genoas (see page 37), it becomes practicable to use different cloth weights for different parts of the sail – lighter panels forward and in the less loaded centre, and heavier panels along the leech and foot to absorb the loads concentrated there. Usually a sacrificial strip is added to the foot and leech to screen the furled sail from UV light.

A final point. Roller reefing genoas should be cut with a highish clew. The more equal in length the foot and leech are, the less you will have to move the genoa lead block forward as you reef. Figure D21.1&2 demonstrates the difference between a high and low cut sail.

In-mast Mainsail Roller Reefing

This works in much the same way as genoa roller reefing. Again the sail is wound onto a vertical spar, housed either in the mast section itself or into a retro-fitted aluminium extrusion riveted to the aft side of the mast. The mainsail is loose footed, and the clew moves forward along a track as the sail is wound in. Such mainsails are usually cut very flat, and maintain their shape acceptably well over the whole reefing range.

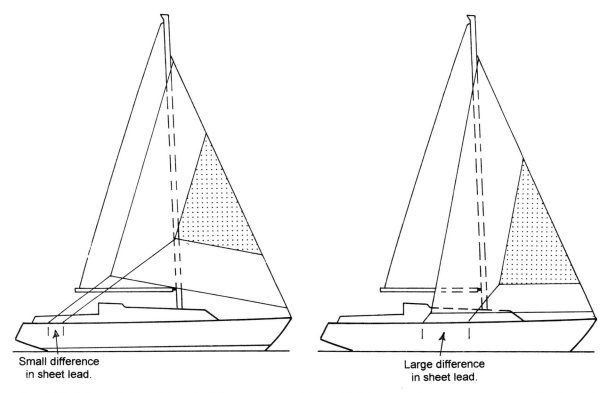

Fig D21.1 Small sheet lead difference   **Fig D21.2** Low cut clew — more adjustment

Although undoubtedly convenient, and probably a sensible choice for the weekender or coastal cruiser, in-mast reefing has some serious shortcomings which could discourage the more serious offshore sailor. For instance:

- Battens are clearly out of the question, so the mainsail must be cut without roach. Including the high cut foot, there will be a significant loss of sail area. But, if the rig was designed from scratch, this could be compensated for in the original sail plan.

- The mechanism adds de-stabilising weight aloft (the retro-fitted units more so than dedicated masts), and this weight remains there even when reefed. Of course, the same can be said of genoa reefing gears, but rather less significantly as these are generally lighter.

- In the event of mechanical failure it may be impossible to lower the mainsail. Unlike genoa roller reefing, you cannot unwind the sail simply by passing it

The Innovative Yacht

round and round the stay. To have it jammed aloft in severe conditions doesn't bear thinking about.

- The mechanism is buried inaccessibly inside the mast. Emergency repairs at sea could be very difficult. If, as I do, you subscribe to Murphy's Law – 'If it can go wrong, it will' – then this could seriously undermine your confidence.

In-boom Mainsail Roller Reefing

This is the up-dated version of the old and trusted (dare one say 'traditional'?) roller reefing, where the main was wound down upon a rotatable boom. In-boom reefing works similarly to in-mast roller reefing, but this time around a horizontal axis, rolling the sail inside an over-sized boom.

This method overcomes some of the disadvantages of in-mast reefing (it accommodates battens – albeit awkwardly, there is no extra weight aloft, and the sail can be lowered come what may), but it must be questioned whether it is a significant improvement on old-fashioned roller reefing. Granted, the control lines are led back to the cockpit, and it all looks very tidy with the sail stowed inside, but for my money there seems a lot of extra cost with very little gain.

And there can be unexpected problems: A friend of mine was doing a stint at the London Boat Show, on the stand of a smallish spar and sail maker, when he was approached by an elderly gentleman interested in in-boom reefing. 'I see you don't make one', he said, peering at the various exhibits.

'Actually', started my friend, winding himself up to the kind of verbal effusion for which he is renowned, 'we don't think they're very satisfactory', and went on at great length to explain why.

'Glad to hear it', said the elderly gent at last when a gap unexpectedly presented itself. 'I have one and I don't like it at all.'

'Aha!' exclaimed friend, and the air was again filled with sales talk.

'It's the sparrows', sighed the gent and succeeded in creating a silence. 'They get into the slot and can't get out. Do you know, by any chance, of anything that'll remove decomposed bird from sailcloth?'

Slab or 'Jiffy' Reefing

Here I must reveal my own preference. This form of reefing gives by far the best sail set, and is simple, reliable, effective, and inexpensive – almost everything you could ask, it seems to me.

A typical arrangement is shown in Figure D22. If the reef is to be handled from the mast, only a single downhaul to each leech cringle is required – the luff being simply tugged down by hand and secured. This arrangement is usually preferred by single-handers (or short-handed crews where only one person might be on deck at a time) because, should the sail snag, they are already in the right position on deck to clear it.

Aloft

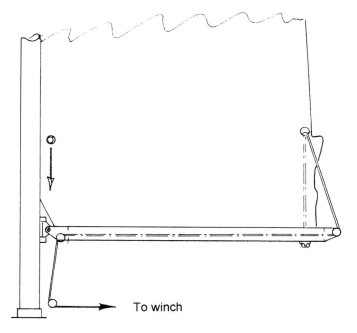

Fig D22 Jiffy (or slab) reefing

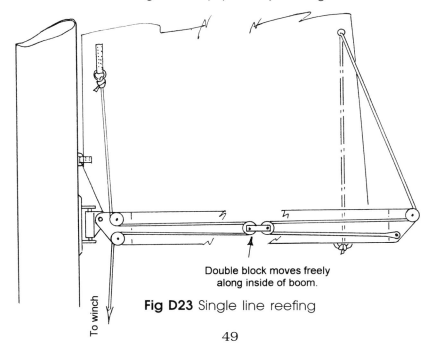

Fig D23 Single line reefing

With larger crews, control from the cockpit becomes more practicable. Now you would need additional pennants to pull down the luff – usually easy to arrange but beware the resultant piles of spaghetti around your feet!

An interesting variant is Single Line Reefing (Fig. D23). Here, tension on the leech downhauls is transmitted via small back-to-back blocks inside the boom. Only one line for each reef point (the luff pennant) is led back to the cockpit. At first sight this would seem the perfect solution, but it does have its problems: Firstly, there is more friction to overcome when re-hoisting the sail and, secondly, there may not be enough distance of travel (for those blocks within the boom) to allow a very deep reef to be made. This leaves you with the prospect of resorting to conventional slab reefing at the very time when conditions are the most severe.

Personally, I would choose slab reefing above all other. But what about the headsails? Well, being a reactionary sort of chap at heart, and instinctively suspicious of all mechanical intricacies, I agonised over this question for months when I was designing *Spook*. But then a trip aboard a friend's boat helped clarify my thoughts. Tacking around Start Point in a rising wind, I had rashly volunteered to go forward to shorten sail. It was late autumn, bitingly cold, and the water seemed determined to invade my oilies. I had performed this task countless times before, of course, but now my middle-aged bones were finding it no fun. Fighting that recalcitrant canvas to the deck, I knew there had to be a better way. By the time I had squelched back to the cockpit to restore my spirits with spirit, I had made up my mind – roller reefing it had to be!

And, having succumbed, I became a total convert. Now I only go on to the foredeck to sunbathe or anchor. And, blissfully, an unforeseen bonus – the cabin is for the first time free of all those damp sail bags.

But, a cautionary word: Buy the best quality gear you can afford, for this is an area where you really do get what you pay for. There is plenty of cheap equipment on the market which might satisfy the occasional sailor, but for serious cruising only the best will do.

Full-length Battens

In an earlier life I designed racing multihulls, and came to appreciate full-length battens before they had gained the acceptance they enjoy today. In that application their purpose was to support the extravagant roach that multihulls favoured (for a given mainsail area, it keeps the Centre of Effort low – an important consideration when capsize is an issue), but other benefits soon became clear.

In the sort of very light breezes when an unsupported sail would tend to collapse, full length battens will maintain the foil shape and help keep the sail driving. A secondary bonus of this generally calming effect is that the sail is less likely to flog, sparing the nerves and also prolonging the life of the sail.

On the debit side, there is obviously more weight aloft, and chafe can occur where the batten pockets rub against rigging when off the wind. The battens are under compression and can slam forward and damage the mast track if the pocket-ends and slides are not properly designed. Smaller mainsails have slides which run on the face of the mast extrusion itself. Larger rigs usually have a special track on which wheeled or recirculating ball type slide cars run.

The battens themselves are usually made of pultruded epoxy/glass section, machined into a taper so that they bend into a foil-like profile, more towards the luff than the leech. This makes them weakest where the loads are greatest and it is not uncommon to see a splintered end make an unwelcome (and expensive) appearance through the leeward face of the batten pocket.

So, full-length battens certainly offer some very real gains but there are also some important losses. It is one of those sailing subjects on which sailors will continue to disagree; and it all boils down to a matter of preference anyway.

Lazy Jacks

This is the sort of contrivance in which we can all delight – cheap, simple to operate, trouble-free and useful. Would that all marine gadgets were the same! Lower the sail and the bunt drops as if into a cradle, ready to be secured or reefed. They work particularly well with fully-battened mainsails, which virtually stack themselves as they descend.

Figure D24 shows the best arrangement. Here the length adjustment is on the boom (some have an uphaul on the mast) where it is both neat and accessible. Sometimes when hoisting the main, the headboard can foul the lazy jacks. It is therefore important, to have the running end long enough to allow the lazy jacks to be slackened off and carried forward until they clear.

In my opinion, lazy jacks are an indispensable accessory to slab reefing – whether single line, double line, operated from the cockpit or at the mast. Without them you must fight that flapping, unmannered sailcloth at just the moment when you have quite enough other things to do. If you are shorthanded they are invaluable; fully crewed and they are still worth having.

Spinnakers

Although it never seems to happen for me, I understand some sailors regularly get to sail downwind. On this point of sailing the boat's speed subtracts from that of the wind, and the resultant 'apparent wind' obviously becomes feebler. In these circumstances there is a need to set more sail than is available from the vessel's working rig.

The true spinnaker is a balloon-like symmetrical sail made of a lightweight nylon cloth. It is set flying from a position above (and forward of) the top of the forestay, and is controlled by sheets and guys from each clew. A pole is usually set to windward to aid in this control.

The Innovative Yacht

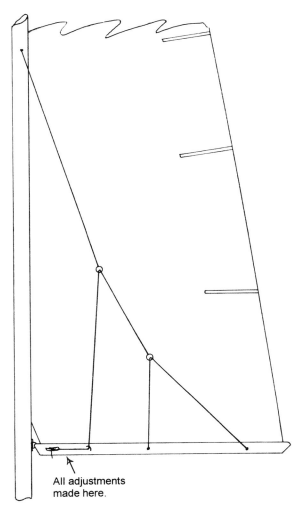

Fig D24 Lazy jacks

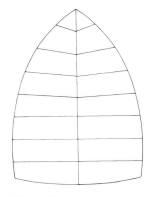

Fig D25.1 Cross cut spinnaker

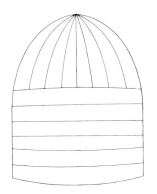

Fig D25.2 Radial head spinnaker

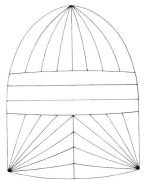

Fig D25.3 Tri-radial spinnaker

Evolution has seen changes in the cut of spinnakers. Figure D25 (1) shows a cross-cut sail of fairly early vintage, (2) is a radial head sail, which is still currently popular, and (3) is a modern tri-radial design for the most demanding usage. On a small to medium sized yacht a radial head spinnaker would be quite satisfactory. But a tri-radial would be a better, though more expensive, all-round choice (especially for the larger yacht), giving flexibility of performance from a dead run to a reach.

Many people are frightened of spinnakers, regarding them as rampant monsters, bent on untold havoc. This is unwarranted. Handled methodically, and with reasonable caution, the spinnaker is a very practical cruising sail. And the sensation you get when running downwind (so I am told) is sublime!

Various devices are marketed to help tame the spinnaker. Going under many names – snuffer, squeezer, sock et cetera – they all work by pulling a fabric tube down over the sail to stifle it. For the more timid kite flyer, one of these could be the answer.

Cruising Chutes

These are asymmetric spinnakers designed to be flown without a pole – though they work much better with one. The tack is attached to the foredeck and the sail is set flying like a rather baggy genoa, controlled by a single sheet from the clew.

The enthusiasm for these sails focuses on the common perception that they are easier to deal with than conventional spinnakers. This is not always the case, and most experienced sailors would counter by claiming that any convenience gained is not worth the loss of sail area and performance. Indeed, in some respects cruising chutes are actually *more* awkward to handle – gybing, for example, can only be accomplished by lowering the sail, altering course, and hoisting it again on the other side (a very laborious exercise), or by passing it around forward of the forestay (which requires very long sheets).

However, cruising chutes will continue to have appeal for short-handed sailors, who might cheerfully trade some loss of performance for peace of mind. Personally, I dislike flying spinnakers at night, but feel perfectly at ease with a cruising chute set. Ideally, of course, a sailboat should carry both.

High Performance Asymmetric Spinnakers

Although there are physical similarities, the true high performance 'asymmetric' is about as closely related to the cruising chute as a jet engine is to a rocket. The most obvious difference is area. The asymmetric is tacked to a pole rigged bowsprit-like forward of the bow, and the clew reaches backwards as far as it can reasonably be sheeted.

The asymmetric is only useful on boats which sail so fast as to be disadvantaged on a directly downwind leg – usually multihulls, or ultra-light displacement craft. In such cases the apparent wind (the difference between boat speed and the wind speed from astern) quickly drops to useless levels and the yacht can sail no faster. Instead, by broad reaching with an asymmetric set, and 'carrying' the apparent wind forward as the boat speed increases, these high speed flyers can scorch along nearly dead downwind, making useful ground over their rivals.

But is a high performance asymmetric spinnaker useful on a cruising boat? Possibly on the very fastest but not really on most. If you can sail no faster than hull speed then the dynamics of downwind sailing are against you. However, the retractable tack pole could very usefully be adopted, and used with a more conventional cruising chute. For it allows the sail to be tacked or gybed more or less like any other headsail, without first lowering it to the deck.

Storm Canvas

The best laid plans sometimes go awry and all of us can expect to be caught out eventually. Our working sails will adequately cope with perhaps ninety-nine per cent of our sailing, but it is that remaining one per cent which contains the greatest peril.

No matter how good the quality, it is unreasonable to expect a deeply reefed mainsail or genoa to function properly in ultimate conditions. This is a job for specialist sails. Every offshore yacht should carry a storm trysail and storm jib.

Ideally, the trysail should have its own track, reaching right to the deck, and should run on slides which can be hanked on whilst still in the bag. To set it on the main mast track means that the mainsail must first be lowered and removed before its track becomes available – an obviously undesirable complication. The trysail should sheet to specially located blocks on deck – not to the boom which should be immobilised to prevent it thrashing around and braining the crew.

In descending order of desirability, the storm jib should be carried:

(a) On the inner forestay, if a cutter (or sloop or ketch with such a stay). Clearly, this brings the sail to where it can be hanked on with maximum security. And the inboard position will also help balance.

(b) On the forestay itself, if no genoa roller reefing gear is fitted.

(c) On a secondary, removable stay set immediately behind the forestay. Many yachts have this fitted anyway – either as a reserve in case the reefing gear fails, or in order to carry a light weather drifter.

(d) Set flying from a convenient anchorage on the foredeck. This is by far the poorest arrangement, but I have used it myself when presented with no alternative, and can confirm that it works after a fashion. Most storm jibs have a wire luff which can be tensioned just about adequately if the halyard is cinched down hard on a winch. If the boat has two fore halyards free – at least one of which must be wire – then the wire halyard can be clipped to the deck and used as an improvised stay, running the sail up with the remaining halyard.

Whatever the arrangement, the important thing is to have it all worked out in advance before severe weather strikes. On many boats I survey, the storm sails still have the sailmaker's original tie about them – having obviously never even been spread, let alone hoisted. A dark night in awful conditions is no time to debate what goes where and how.

Spars and Rigging

Where once varnished spruce held sway, the ubiquitous tin stick now rules – and with very good reason. A light, strong metal offering excellent corrosion resistance, which can be extruded into thin-walled hollow shapes, aluminium alloy has to be a winner for building masts, booms, and spinnaker poles. Toss in abundant supply and reasonable cost and it seems almost too good to be true.

Early aluminium spars relied heavily on welding for their manufacture. Masts were often tapered, and such features as masthead cranes, winch bases, and sheave boxes were entirely integrated with the basic section. But, today, most modern masts are an assemblage of modular fittings which are rivetted or bolted together by largely semi-skilled labour. There are obvious cost advantages in this approach and, in this regard, the sailor is the partial beneficiary.

But not all of these developments are welcome. Too often strength and reliability takes second place to manufacturing convenience. Is a cast alloy goose neck really better than one fabricated from stainless-steel? Hardly. Will an injection-moulded plastic exit sheave last as long as one made from aluminium? Not if exposed to tropical sunlight, it won't. These economies might give the mast builder an edge in the marketplace but they certainly do no favours to the sailor far from land.

Problems rarely arise from the mast sections themselves. Compression loads are readily computable and, except on very light rigs where engineering margins have been shaved to the minimum, failure should not occur. It is the rigging or fittings which usually let us down.

Dismasting is a major sailing catastrophe and should be avoided at all costs – perhaps quite literally by not going for the cheapest options. And it's not as if the extra expense need be very great – indeed, it should be inconsequential compared with those you would face if it all came tumbling down.

Of course, if you buy a stock or second-hand boat, you may have to choose between living with the mast that comes with it or replacing or modifying it. But, if you have any choice in the matter, you might like to consider the following points:

- Conventional tangs and eyes are the most forgiving method of attaching shrouds. T-terminals and stem-balls are wretched devices, designed to save money not masts. Remember that all standing rigging must have sufficient articulation (at both ends) to accommodate the natural flexing that will occur. If misaligned or restricted in movement, stainless-steel wire will fatigue and fail.

- Masthead assemblies fabricated from welded plate or extrusions are less treacherous than alloy castings. There may be voids or other faults in a casting which are invisible from the outside.

- Cast goose neck fittings – especially those where aluminium grinds directly against aluminium – are always suspect (though at least at deck level you have a fair chance of spotting problems before they become severe). The critical components should be of stainless-steel.

- Other important items such as spreader sockets and kicking strap anchorages are also better fabricated than cast. Remember that the loads can be awesome, and that the original strength of a fitting can be seriously reduced by wear.

Rigging Screws

Often called 'turnbuckles' or 'bottlescrews'. These are the tensioning devices for the standing rigging and, obviously, are critical components in the overall security of the rig.

The vast majority are machined entirely from stainless-steel. This is not as good a choice of material as it might at first seem. Two stainless-steel surfaces sliding over each other under high loads can gall and fuse together. This problem can be overcome by using dissimilar metals. The best rigging screws have threaded stainless-steel studs running into bronze bodies – or sometimes stainless-steel bodies with bronze inserts. If you are looking for trouble-free service from your rigging screws, these are the only kind to have.

And respect the humble toggle. It is as vital to articulate the bottom end of a stay as it is the top. Many rigging screws have a toggle built in – an elegant and reliable approach to the matter. But, whether integral or a separate unit, a toggle is essential on every rigging screw, no matter how well the chain plate is aligned.

But if the toggle is wholly benign, then the split ring is a minuscule menace. Often used to secure clevis pins, these evil little devices can be snagged by flogging ropes (usually the foresheets) and pulled out. I have known of at least four masts (one of which was mine) that fell down through such a cause. Now I won't have them anywhere near our boat. Split pins (clevis pins in the US) of the correct size must always be used. The bifurcated ends should be opened out to about 60° (never turned back on themselves) and taped over to protect sails and passing ankles.

Chapter 4
On Deck

The deck is the working platform for any boat. This is not to say that it cannot also do service for the alluring (or otherwise) disportation of bronzing bodies, or for jollying oneself over oysters and a few bottles of Muscadet, but its primary function is to provide a convenient and secure base from which to handle the ship.

This rather elementary point is overlooked by some yacht designers who, no doubt seduced by the jelly-mould potential of GRP construction techniques, set out to fashion their decks with hardly a thought to the poor souls who will have to move about on them – and, to boot, move about on them when the darned things are sloped at crazy angles. Take a stroll down any marina and cast a critical eye over a few decks. All sorts of swoops, hollows, and other seemingly functionless obstacles will been found lying in wait for the unwary. Some decks are almost as perilous in the dark as the north face of the Eiger. Of course, different parts of the deck must satisfy different requirements and each must be assessed accordingly. Let's take a look at the various areas in turn.

Cockpit

The command centre of your yacht, and the area where you will spend the most time at sea. The design of the cockpit will be crucial to your ease and security. It is both your treadmill and sanctuary – where you will toil and shelter. On long trips it will come to seem like your home.

No feature more plainly identifies the region for which a yacht is designed. Cockpits in high latitude boats are small and deep – foxholes in which the crew can cower. Those for sunnier seas will be as open and spacious as beachfront patios; a place in which to languish rather than hide. The long-distance sailor, straying between meteorological zones, will strike some sort of compromise. Form should follow function, as always.

The Innovative Yacht

Fig D26 Deep cockpit

Fig D27 Properly angled coamings

Cockpits should be as deep as is practicable – though the modern fashion for interconnected aft cabins, with their attendant need for passageways running beneath the cockpit, often limits this. Not only will depth give better shelter, but it will place the winches at a more comfortable level to work (Fig.D26). Careful, though: A deep cockpit increases floodable volume, and the drains must be adequate to cope. Also, visibility may be restricted, obliging the watchkeeper to stand up frequently to keep a weather eye out for traffic. Once, when crossing the Irish Sea, I awoke to the thumping beat of a ship's engine and scrambled on deck to see a tanker cross our bow. The helmsman, settled deep in the glassfibre canyon on this particular boat, had been oblivious of its approach.

Although most cockpit seats are tolerable with the vessel level, some can become almost untenable when heeled. If the coamings are inclined as, say, the back of a dining chair might be, then the windward one will be too upright, or may even lean inboard, with the boat hard pressed. And the deeper the coaming, the more pronounced this effect will be – to the point where it can become virtually impossible to stay seated. Well designed coamings should be properly angled to make the weather side comfortable when sailing (Fig.D27). And, ideally, the seats themselves should be profiled to support the back of the thighs without any hard corners digging in. But this must be done with moderation. Extravagantly radiused edges can make deceptive footing when you are leaping around in the dark.

On Deck

Because the cockpit is not as potent a sales feature as the interior, it often receives less attention than it deserves from the designers. Worse still, its layout may even be compromised by the temptation to squeeze in some other more marketable facility below. But cockpit comfort is not a trivial matter. To be forced to sit awkwardly will inevitably accelerate fatigue. Physical misery can quickly erode crew morale, perhaps to the point where inattention and carelessness could occur.

Drainage

If flooded, perhaps by pooping, a typical cockpit-well could easily hold half a ton or so of sea-water. And this quantity would take over a minute to empty through a pair of 1.5in (38mm) diameter drains, and double that time if these were reduced to 1in (25mm) diameter. Either way, it could be quite long enough for the now hampered vessel to be swamped by another breaking sea.

Drains should always be as large as practicable and designed without constrictions. To have a large diameter hose choked-off with a smaller fitting somewhere along its run would be to waste much of its potential efficacy. Similarly, to have multiple drains siamesed into a single seacock might look impressive from the outside but would perform no better than a single drain. Seacocks should be of the full flow type, and properly matched to the hose diameters.

Generally speaking, gear is scaled to the size of the yacht – big boats have big winches, and so on. But the reverse should be the case when talking of cockpit drains. Half a ton of water sloshing around the stern of a medium displacement fifty-footer would be bad enough, but imagine it pinning down a compact family cruiser.

Transom freeing ports make very effective drains. But it is important that these should be equipped with one-way flaps (Fig.D28) to prevent following seas surging into the cockpit.

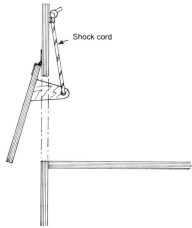

Fig D28 Transom freeing port

The Innovative Yacht

Wheel or Tiller?

The ruination of many a good aft cockpit is the choice of wheel steering where a tiller would do nicely. Centre cockpit boats, with their need to route the controls through the accommodation, clearly have no alternative.

Conducting a thoroughly unscientific straw poll amongst friends, I concluded that preferences depended largely upon whether a person started sailing before or after he learned to drive a car. Latecomers often felt more familiar with a wheel; those who sailed dinghies as youngsters were more ready to go for a tiller.

In small to medium sized yachts there is very little to be said in favour of wheel steering where it can be avoided. It is more complicated, troublesome, and expensive; it places the helmsman further aft where there is least protection; it is less precise in operation than a tiller; it clutters up the cockpit and cannot be hinged up or removed when in port; it makes connection to both vane and electronic self-steering gears more awkward; and it adds extra weight aft where you least need it.

As boat size increases, the benefits of wheel steering become greater and the penalties less: The heavier steering loads can be reduced by mechanical gearing (rather than resorting to extraordinarily long tillers); cockpits become large enough to accommodate the steering pedestal; auto-pilots are usually integrated units, acting directly on the rudder stock; and the extra weight and costs become relatively insignificant parts of the whole.

Commonly, manufacturers will offer wheel steering as an expensive 'extra', simply installing the mechanism without any modifications to the seating. Rarely will a boat designed for a tiller be so instantly adaptable, and the helmsman may find himself adopting all sorts of contortions in order to steer. If wheel steering is to be the choice, then the cockpit should be specifically designed for this purpose, with a properly shaped helmsman's seat which is comfortable at any angle of heel.

Fore and Side Decks

She was a brand new aluminium ketch of French build, as flush decked as an aircraft carrier, save for a tiny doghouse.

'Just look at all that open deck space' said the owner proudly. 'You could hold a village fête on it – with room to spare for the morris dancers. And the sunbathing . . . !' His lower lip trembled as his thoughts turned to nubile bodies, Ambre Solaire, and his own role as applicator.

That evening we put to sea, in no danger of any sunbathing. I took the helm whilst the owner bustled forward to secure the anchor. With main and mizzen set, we sailed clear of the estuary. The boat leaned away from the breeze. Rain rattled on the deck. The owner finished his work and made to return. Then, on his face I saw contentment freeze to consternation. The horizontal promenade he had previously admired was now a glistening inclined plane with not a decent handhold in sight – a no-man's land of exposure between himself and the cockpit.

For a deck lay-out is more than just walkways. Thought should be given to every task that may be performed there; handholds for each purpose must be provided. These may be guardwires, grab-rails, or even the standing rigging, but there needs to be a reachable succession of them from one end of the boat to the other. Going forward in heavy conditions is something like playing Tarzan swinging through the jungle canopy.

All non-slip surfaces should be just that – non-slip. The moulded texture on many GRP decks is often disappointing. None are very good; some are downright glacial. By contrast, the stick-on cork compound surfaces – such as Treadmaster™ – are excellent, though comparatively expensive. Gritty non-slip deck paints are a cheaper, acceptable alternative, and have the added benefit of being easily touched-up to repair minor damage.

Contrary to popular belief, teak decks have rather poor slip resistance, especially when wet and if bare footed. Though undoubtedly attractive, and gentle to skipperly sterns when sat upon, teak planking can be a menace on decks and cabin tops. It also adds weight to your hull at a point disadvantageous to its centre of gravity.

But whatever non-slip system is used, it is essential that all surfaces on which you might stand will be completely covered. Watch out for radiused corners to deck houses and coamings. At the behest of our old foe, 'styling', these are usually left glossy – hardly a sensible practice. When you remember that to slip could be to fall overboard and drown, the choice between safety and vanity becomes a less difficult decision to make.

Jackstays

But, despite all precautions, the unthinkable could happen to any of us someday. A lurch of the boat, a moment of unguardedness, and there we are over the side. Too late for precautions now. We must rely on a harness to keep us tethered to the boat.

I have this vision of the men who design harnesses. They are the sort who compile cryptic crosswords, or devise idiotic game shows for television. Some harnesses are as baffling as Chinese puzzles. And I knew one man who actually dislocated a shoulder squirming into his. Amongst the essential qualities of any harness is that it should be easy to put on.

Personally, I carry two – one adjusted to fit over my oilskin jacket, and another which fits me stripped or nearly so. Many sailors favour the kind of harness built into a jacket, which goes on automatically when you are kitting out for heavy weather. This might be a good choice in higher latitudes but would be far too cumbersome in the tropics. The attachment lines should be fitted with the type of carbine hooks which cannot be accidentally tripped.

The jackstays should be rigged so that you can clip on before leaving the cockpit. The most popular wire type will roll underfoot (perhaps precipitating exactly the type of accident it is designed to preserve you from) so I much prefer them to be made of strong polyester webbing (1in (25mm) wide, 1,800kg breaking load) which lies flat.

The Innovative Yacht

And, rather than leave them on deck permanently where UV light and trampling feet can attack them, I have soft eyes sewn into each end so they can be rigged or put below in a matter of minutes. The forward eyes are long enough to be cow-hitched through the mooring cleats; and the aft eyes carry 6mm-diameter rope lanyards which are lashed to convenient strong points.

Pulpits, Stanchions and Guardwires

These are the safety fences of sailing which, hopefully, will prevent you falling into the drink in the first place.

First, it is essential that they be high enough – 24in (61cm) should be the absolute minimum, 28in would be better. And there should be no reduction for smaller boats, where on-deck security is of greater importance, not less. Ideally, at least the top guardwires should be covered in white PVC. In poor light this makes them easier to see and grab.

The leverage caused by a heavy man falling against a stanchion imposes formidable loads on the base. Substantial backing pads beneath the deck are a must – but are frequently omitted by penny-pinching builders. Considering the importance of these fittings, it is well worth taking the trouble to check them out for yourself. The costs of improving the installation could be modest – the consequences of failure, tragic.

Winches

Following an address to a yacht club gathering, an eminent old single-hander was asked which development had been the most useful to him in his numerous ocean passages. Without hesitation he replied: 'Self-tailing winches'.

This took the audience by surprise. What about GPS, they enquired, and autopilots and solar panels and radar?

'You won't get a hernia operating a sextant', he replied gruffly.

Whereas the full racing crew has a surplus of crew available to tail, the average cruising sailor often finds himself struggling with the sheets on his own. For him the self-tailing winch is an unqualified boon. Not only can he take both hands to his work, but he can also position his body in such as way as to get the best purchase.

The simplest winches are single-speed, direct drive devices, where the mechanical advantage is derived from the difference between the radius of the drum and the length of the handle (Fig.D29). It can be seen that to buy a larger winch (thereby going up in drum diameter) will actually give you less mechanical advantage, not more – a point often overlooked by potential purchasers.

The direct-drive winch clearly has its limitations. If more power is required, internal gearing must be employed. The smallest geared winches are invariably single-speed; mid-sized winches are usually two-speed; and the largest often three-speed.

The right winch for each job is dictated by the demands of each specific task, perhaps adjusted for the strength of the crew. Power is gained at the expense of line speed (a winch having a power ratio of 50:1 will wind in only half as fast as one of 25:1), so

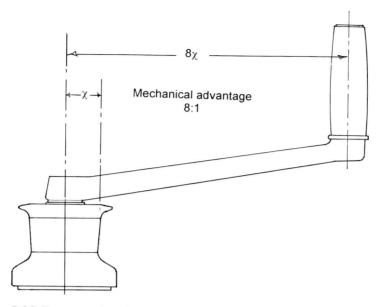

Fig D29 The mechanical advantage of a single speed winch

to choose a needlessly powerful winch can be counter-productive. The table below shows a sensible compromise between power and line speed for a variety of functions aboard boats of different length. The power ratios shown are for the lowest, most powerful reduction using a 10in (25.4cm) handle. Fractional rig sailboats may use genoa winches one size smaller. Light or elderly crews might choose to go up a notch.

RECOMMENDED WINCH SIZES

LOA (feet)	LOA (mtr)	Sail Area (sq ft/sq m)			Genoa Sheet Cruising	Spinnaker Sheet Cruising	Main Sheet Cruising	Genoa Halyard Cruising	Spinnaker Halyard Cruising	Main Halyard Cruising	Staysail Halyard Cruising
		Genoa	Spinnaker	Main							
25–29	7.6–8.8	300/28	400/37	150/14	16/24	7	6	7	7	6	6/7
29–33	8.8–10.1	350/33	600/56	180/17	30	8	6	16	7	7	8
33–35	10.1–10.7	470/44	800/74	210/20	40	16/24	7	16/24	8	8	8
35–37	10.7–11.3	550/51	1000/93	230/21	44	30	16	30	16	16	16
37–39	11.3–11.9	600/56	1200/111	260/24	48	40	16	40/44	16/24	24	16
39–41	11.9–12.5	750/70	1400/130	300/28	50/54	44	24	44	30	30	24/30
42–48	12.8–14.6	900/84	1600/149	350/33	58/62	48	40	48	44	40	40
48–55	14.6–16.8	1500/139	3000/279	750/70	66	54	48	50	48	44	44
55–62	16.8–18.9	1900/177	3800/353	875/81	77	62	54	54	50	48	48
62–71	18.9–21.6	2300/214	4600/427	1000/93	88	66	62	58	58	54	54

Anchors, Warps and Windlasses

The ground-tackle can be thought of as the fingertips of yachts; and when you need them to hang on, they had better be good and strong.

Amongst anchor manufacturers there is a curious inversion of marketing philosophy. Instead of trying to sell you the largest and most expensive anchor they can, they seem to go to considerable lengths to persuade you that smaller will do. Each has incontrovertible data that their particular baby will handsomely out-perform all the others, and will wave the evidence before your eyes like a host of biblical prophets recently come into stone tablets.

By individual sailors, anchors can also be regarded like magic nostrums. 'I'm a Portland Pick man, myself. Rode out hurricane Charlene on nothing more than a fifteen-pounder. Wouldn't trade it for a truck full of . . . .' You know the sort of thing.

The truth is that all modern anchors are pretty good, but some cling better to certain bottom types than others. An anchor that works well in soft mud may be less efficient in sand or weed, and vice-versa. It is sensible to carry a selection – three at least for serious cruising, all of adequate size. And foreknowledge of the conditions you might meet can be invaluable. Preparing for a Round Britain Race, I was told that the kelp in Castle Bay (the second stop) made the holding treacherous. Anticipating this, we threw a traditional fisherman type anchor into the forepeak and, in the event, rode to it snugly whilst other competitors dragged around us.

And there is no excuse for carrying anchors which are too light for the job. With a decent windlass and properly designed self-stowage, the appropriate anchor for the size of vessel should be easily handled. The popular 'lunch hook' – typically a puny thing, looking as if it fell off a charm bracelet – has no place on any sailboat. Even if anchoring only for a few minutes, tethering yourself securely to the bottom is a worthwhile investment in peace of mind.

Windlasses

Taking an early morning row through a Caribbean anchorage, I was hailed from a nearby ketch. A few strokes later I was alongside. A slender grey-haired lady took the dinghy painter and bade me step aboard. Her husband grinned sheepishly at me from the cockpit.

'Wondered if you'd mind, old chap. Put my back out again and the wife can't raise the bloody anchor. Should have married a Russian shot-putter, of course, but Maud does mix a wondrous rum punch. If you could just get us under way, we can make it back to St. Thomas.'

Which I did, weighing anchor and hoisting their mainsail before climbing back into my dinghy.

The moral of this story is obvious. When sailing shorthanded, a sailboat must be equipped so that routine functions can be carried out by any member of the crew. To be entirely reliant upon the strongest to do the heavy work leaves you very vulnerable.

There comes a point where large sailboats must fit electric or hydraulic windlasses, so that takes care of that. But for small to mid-sized yachts the manual windlass (if any) is a viable option; less costly in both cash and current drain than the powered variety. Of these, there are two distinct types:

- The horizontal-axis type operated by cranking a handle back and forth. These are the most powerful units, suitable for boats of up to about 55ft (16.8m). Internal gearing gives high pulling power, but chain recovery speeds are slow. Often, a warping drum is incorporated.

- The vertical-axis capstan type, suitable for sailboats up to about 40ft (12.2m). These are operated with a standard winch handle, and there is no internal gearing to amplify mechanical advantage. Extremely simple and robust, the vertical axis windlass winds in fast, but is shorter on power than the more elaborate units.

The smaller windlasses often have gipsies which will handle warps made up of both rope and chain. Many can be upgraded to electric drive with the addition of retro-fitted motors, installed below decks. Except for the very largest sizes, all hydraulic and electric windlasses can be operated manually in emergencies.

In combination with self-stowing bow fittings, a useful refinement on electric installations – especially those with a powered reverse facility – is to have a set of controls in the cockpit. This makes leaving a crowded anchorage a wonderfully nonchalant procedure, and guarantees total co-ordination between cockpit and foredeck when backing into a Mediterranean-style marina.

The 'Mediterranean' Moor
The classically convivial way of parking your boat in the Med. Round up some thirty or so metres from your chosen slot, drop the hook into the hoggin, and with maximum aplomb slide backwards in towards the quay, hopping ashore with the stern lines at the last moment. Done well it looks masterly. Done badly it doesn't.

Usually, I prefer to do it the other way round – bow-in, that is. Not only is it a less chancy manoeuvre but, when all secured, there is also more privacy from the gawping hordes trailing past. Perhaps not surprisingly, in Mediterranean countries the locals find marinas amusing places to perambulate. Once, following a lunchtime run ashore, I stumbled up from my siesta, stupefied and naked, and was halfway into the cockpit before I remembered where I was. The hilarity of the three little girls standing on the quayside was only exceeded by the outrage of their mother.

But the bow-in moor obviously requires an anchor astern. And, if you are planning to use it regularly, some permanent arrangement would be convenient.

The Innovative Yacht

Hinge-down pulpit step for boarding over the bow.

Portable pulpit-mounted boarding ladder, stowed below when not in use.

Stepping Ashore

By design, sailboats are difficult things to get off; a virtue at sea but a darned nuisance in port. If alongside a pontoon, to swing ones legs over the guardwires is not too demanding – though an opening gangway will make even that a good deal easier – but, when moored bow or stern-to, clambering ashore can call for almost simian agility.

Walking the plank is the answer. The passarelle should be a part of every Mediterranean cruising inventory. The inboard end is secured to the stern (often in a dedicated pivot), and the other end either scrapes along the quay or is held clear a few inches above it by an elasticated bridle. The latter is obviously the preferable arrangement, though I do recall inflicting serious damage to a Spanish cat, who was dozing in its shade when we all clattered ashore. Incidentally, a very handy passarelle can be made up from an aluminium ladder to which a plywood plank has been lashed. In their own right, ladders are useful bits of kit to have aboard, too.

If disembarking over the bow, a specially designed pulpit incorporating a hinge-down extension, will greatly facilitate access. With careful adjustment of the mooring lines, the bow can usually be made to overhang the quay.

On Deck

Berthed stern-to — the 'Mediterranean moor'. The *passarelle* topping-lift bridle is made of shock-cord, trimmed to hold it just clear of the jetty. A person's weight depresses it for use.

A small pulpit mounted abaft the headsail furling gear.
The sail sheets outside the pulpit, thereby reducing chafe.

Chapter 5
Come Rain or Shine

It is another of the many ironies of sailing that, having wittingly exposed ourselves to the elements, we then must do our darnedest to hide from them. Of course, there are some rugged souls who like nothing better than having a briny deluge lash their cheeks, or to be barbecued to a crisp under a tropical sun. But most of us are cast more in the shrinking violet mould, cowering from the harsher aspects of the weather.

A friend of mine recently traded his Half-Tonner for a motor-sailer, bloated of hull and meanly rigged. Such a craven defection from the competitive brawling of the club racing scene was received with some astonishment by his mates. But he withstood their scorn with serene good humour, and agreed to join them in Cowes, at the end of the next day's event.

Well, it rained the next day and blew gustily from the West – the sort of weather sailmakers pray for. Battered and bedraggled, the fleet rounded the course in a fury of spray, and belted towards the finish. Just to leeward of the line was the motor-sailer, with only headsail and mizzen set, loping back and forth at an easy pace. As each boat crossed, a shirt-sleeved figure emerged from the wheelhouse, drew nonchalantly on his pipe, and raised a mocking gin and tonic in salute. As all later agreed, he had certainly made his point.

But many would also agree that his was a rather desperate solution. Any amount of protection is possible if you are prepared to compromise sailing quality, but speed and windwardness are not things to be tossed away lightly. Of course, large yachts can carry substantial deck structures without becoming overly hampered, but all smaller craft must be more mindful of extra weight and windage – especially if placed so high. To provide good crew shelter whilst retaining a pleasing, seakindly shape is just another dilemma the yacht designer must face.

Some don't try very hard. Indeed, the most conventional approach is simply to design an open cockpit with low coamings and a hatchway leading downwards to the accommodation. And, as this arrangement is so common, many yachtsmen accept it as being entirely normal, and go along with what little shelter it offers without complaint.

Discomfort and fatigue are two heads of the same black monster, which can lead even the stoutest soul towards debilitation and despair. All sailors, if they stick at it long enough, will experience that 'death, where is thy sting?' feeling, and will know how hopelessly a man can sink when exhausted and dispirited. I vividly recall an incident in the 1970 Round Britain Race when, at last at anchor at the end of a particularly gruelling stage, we were rammed by another competitor coming in. I was almost asleep at the time – or, more accurately, in a state of near collapse, gratefully choosing oblivion over any other need. Jolted awake by the impact, I staggered on deck to inspect the damage. Hardly anything – a loosened cleat and some scuffed paintwork. Where I would normally have felt relief, I then almost wept with disappointment. If he had done the job properly and struck hard enough to hole us, we could have withdrawn from the whole awful affair with honour.

It is my belief that to attempt to keep dry or warm or cool – or whatever is appropriate to the prevailing situation – is not some wimpish indulgence of which we should be ashamed, but an absolute duty. To preserve our strength and spirits by looking after ourselves as best we can is simply common sense. At both extremes – frostbite or heatstroke – the weather can do us great harm, and we need to plan our defences carefully.

Let's consider a few options.

Wheelhouses

The very best protection is afforded by total enclosure. On super-yachts wheelhouses can be lavish air-conditioned capsules, as comfortable and convenient as the bridge on an ocean liner. Naturally, on such exalted vessels sail handling is invariably operated by hydraulic furling gears and captive-reel sheet management systems, all push-button controlled from a central console. The need for any manual exertion is removed, and any undignified scurrying around can be accomplished by lackeys dispatched for that purpose. But, on less exotic sailboats, wheelhouses usually have access to an open cockpit from where the sails are adjusted in the conventional manner.

Although undoubtedly cosy, this 'indoors' approach to sailing brings its own problems. Although most sailboats also have an external steering position, this is usually located behind the wheelhouse where there is probably inadequate seating and an obstructed forward view. Very often this second helm is considered 'secondary', with nothing like as much thought being paid to its design. Except perhaps for short periods, this more or less condemns the helmsman to being stuck inside, no matter how nice it might be on deck.

And, when inside, the actual structure can insulate him from many of the subtle inputs he needs to sail the boat well. No longer can he feel the wind in his hair, or sense the rising note of an approaching squall. A luffing mainsail or other defect in sail trim might go unnoticed above his head, invisible through an opaque coachroof top. Even keeping a sharp look-out becomes more difficult through spray smeared glass – especially at night above the glow of the compass and instrument displays.

But at least those instruments will be close at hand – and this is one of the wheelhouse's strongest claims for favour. Unlike the open cockpit boat where the steering and sail handling are done on deck, with the relatively delicate navigational equipment kept well out of harm's way below, the sheltered wheelhouse can be a complete command centre, with all navigational aids instantly available to the watchkeeper. Anyone who has diced through the traffic lanes in the English Channel will know how comforting it is to have the radar conveniently in view.

Deck Saloons

These have recently become an increasingly fashionable variant of the wheelhouse, embodying many of its features without being dedicated solely to the task of navigation. Here, the steering position is incorporated into the accommodation, typically sharing the same space with the dining area and galley. In order to provide all-round vision, the midlength area of the cabin sole must obviously be raised. The result is an airy cabin, where social and operational functions can mingle under cover – at first thought, a wonderfully attractive concept.

But, despite its attractions, I think this is a misconceived arrangement, more suited to power-boat design, from whence it was borrowed. Clearly, the helmsman must be located towards the forward end of the saloon, so that he can see ahead over the bow. This places him at some distance from the aft cockpit with its sheets and winches – hardly convenient if short handed, and even less so if he is tramping back and forth in his dripping oilies. Of course, if there is the usual alternative cockpit steering position, he can be banished there when the wind pipes up, but this rather defeats the object of having an inside helm at all.

Quite recently, I was asked by a potential client to sketch out a few ideas for a deck saloon lay-out. My initial doodling produced a plan roughly as shown in Figure D30.1. Quite practical, one would think, and the sectional drawing D30.2 seems to confirm this. But now imagine the same yacht heeled over at 30° on starboard tack. The dotted line shows the inclined waterline. If you wish, rotate the page to simulate. The helmsman in his perch to windward (often a motor-boat type pedestal seat) is now clinging on like a climber on a cliff face. What is more, he can now see nothing but sea out of the port windows and precious little but sky to starboard – not a lot of point in him staying there, really.

Having so far been rather unenthusiastic about the deck saloon, I must now confess that my own preference is for something very like it. It occurred to me some years ago that all-round vision from below was a very desirable feature on any offshore cruising yacht. Many people sail short-handed – husband and wife being a typical crew – which means that watchkeeping is often a solitary business. With a yacht properly

Come Rain or Shine

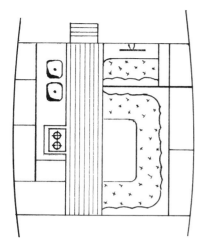

Fig D30.1 Deck saloon in plan

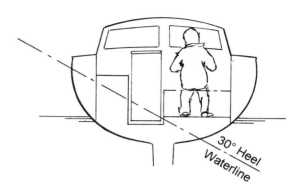

Fig D30.2 — and heeled

set up to self-steer (much more on this later) the watchkeeper is liberated from the tedium of maintaining a course, and free to attend to other, more absorbing tasks.

On *Spook*, our custom-built 31ft (9.5m) sloop, the companionway leads directly down to an area containing the chart table and galley, located beneath a low coach house having virtually wall-to-wall visibility through small toughened glass windows. There is no direct steering, nor sail adjustment possible from there, but remote controls to the auto-pilot and vane gear permit small variations in course – often quite enough to dodge another boat. Cooking, navigation, or simply keeping warm can all be accomplished without the fear that you are just about to sail under the bow of a supertanker. I have spent many hours comfortably wedged down there, with my oilskins ready to slip on in case I must scramble on deck.

The Dog House

Here we have a versatile and seamanlike solution to the problem, especially well suited to the wheel-steered, mid-sized or larger yacht. Despite being open from the stern, dog houses offer very useful protection – though obviously falling short of the cocooned shirt-sleeve luxury of the wheelhouse. The fiercest winds usually come over the bow or the beam, anyway. To tuck yourself well forward under the dog house can make a rough watch tolerable.

The permanent structure is in many ways similar to the folding spray hood, but is obviously more substantial. To improve their effectiveness, they can easily be extended with canvas covers and screens (Fig.D31). In warm climates, it is handy to have the forward windows openable to improve ventilation and visibility.

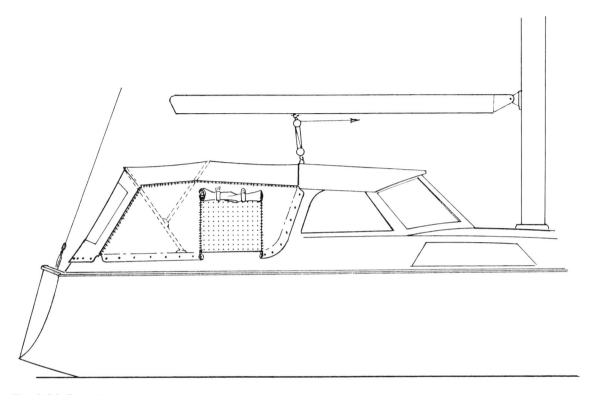

Fig D31 Dog house screens

Windscreens

About as minimal as you can get in rigid cockpit protection, and almost always supplemented by a removable soft top. Ideally, the screens should be of toughened or laminated glass – not acrylic or polycarbonate which will eventually scratch, craze or (in the case of polycarbonate) fog. Again, it is useful if the front screens open.

Spray Hoods and Dodgers

An unprotected open cockpit can be transformed by the addition of portable fabric hoods and screens. The fabric is usually woven acrylic or reinforced PVC, both of which have good resistance to weathering. Most spray hoods (and some dodgers) have flexible plastic windows for visibility – though the view is usually distorted by the poor optical properties of the material.

The folding spray hood is a common sight on cruising yachts – indeed, few open cockpit boats would venture to sea without one. Designs vary, but it is essential that they be made as stoutly as practicable. Not only can they take an awful beating from any waves sweeping the deck, but people tend to cling to them as they negotiate their way forward. In this regard, stainless-steel frames are preferable to aluminium. And, once more, to be able to open up the forward windows is an inestimable boon.

At their most basic, cockpit dodgers are simple rectangles of fabric. This could be an opportunity missed. Pockets sewn onto the inside face make excellent cockpit tidies; loops on the outside can carry lifebuoys. In any event, the addition of the yacht's name will greatly assist identification in emergencies.

So far this chapter has been pitched pessimistically towards the soggier end of the sailing experience, but we all see sunshine sometimes, and in some parts of the world we might even see too much! This is probably astonishing news to some Brits, but it's perfectly true just the same.

Too often, higher latitude sailors stray nearer to the equator without fully understanding what they are getting into. In their minds, sunshine is to be relished and is therefore benign. This is a delusion. In many respects, intense heat is more hostile than rain. With the right clothing, you can keep warm in almost any conditions, but remaining cool when it really gets steamy is another matter. There is a limit to how much you can strip off and, beyond a certain point (and ignoring decorum), even that can be counter-productive. In the early summer of 1974 I sailed to the West Indies in a boat I had originally built for that year's Round Britain Race. Now, the design criteria for latitudes 60°N and 15°N are climatic worlds apart. My trimaran *Whisky Jack* was a foul-weather projectile, in no ways ideal for the tropics. But, my plans had changed and I was on my way to Texas, and it was the only boat I had.

Heading west, the morning sun rose from dead astern and passed directly above us. In the cockpit the forenoon heat was ferocious. There was not a scrap of shade on deck. Fortunately, under twin headsails, the boat was sailing itself. We hid below and let her get on with it. But later, by the afternoon, the sun would have slid forward, and the sails would cast delicious shadows over the whole boat; whereupon we could emerge blinking from the cabin like infantrymen sensing a truce.

The principal requirement, of course, is shade. Block the sun's rays and – assuming there is any, which is usually the case offshore – the wind will keep life cool. Now we are talking awnings.

The Bimini

Named after the westernmost island in the Bahamas, and thoroughly appropriate to the climate there. A bimini (Fig.D32) is a folding canopy, stretched over a tubular metal frame, and braced with straps or lanyards. It can be rigged or stowed in a trice.

The Innovative Yacht

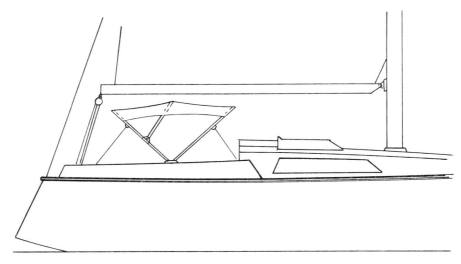

Fig D32 The Bimini

Originating in the United States, they are extremely popular in tropical and subtropical regions but rarely seen in northern Europe. This is a pity because, although intended as parasols, they make effective umbrellas too. I commend them thoroughly to yachtsmen everywhere.

Deck Awnings

Whether a simple boom tent (Fig D33) or a full-length, customised canopy, an awning will effectively extend your accommodation, and will keep most of the direct sunlight off the deck, reducing the temperature below. In warm climates, life is conducted mainly on deck, anyway, the cabin being thought of as a sort of cellar beneath your feet.

Woven acrylic cloth or natural canvas is the preferred material here, and preferably of a light colour, but never blue. This traditionally nautical colour absorbs the hotter red end of the light spectrum and reflects the cooler blue shades – which, of course, is what you see. In this respect, our eyes betray us. Blue looks cooler than red but isn't. White or near-white is the best choice. When the sun is low, side panels (Fig D34) will increase the area of cast shade without extending the awning beyond the beam of the boat.

Awnings come in as many sizes and shapes as there are different boats. It is outside the scope of this book to suggest specific designs, which should reflect individual requirements. But, in the absence of anything more elaborate, a square of canvas, each side being about equal to the boat's beam, and eyeleted along its edges, is a compact and versatile bit of gear which can be rigged in a variety of useful ways.

Come Rain or Shine

Full-length awnings keep the boat cool and protect the brightwork.
However, on smaller boats such as this, they can severely restrict access on deck.

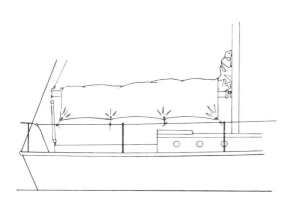

Fig D33 Deck awning

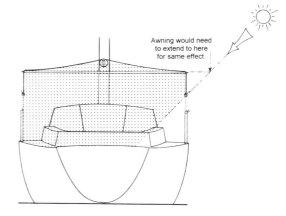

Fig D34 Awning side-panels

The Innovative Yacht

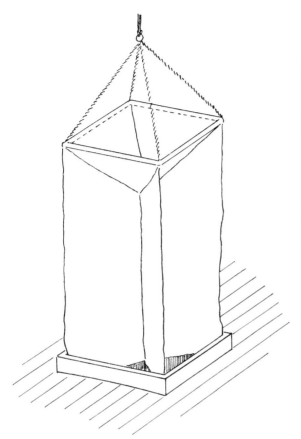

Fig D35 Windscoop

The simplest form of windscoop, drawing nicely and ventilating the cabin

Windscoops

Poor man's air-conditioning, and worth every penny of the modest investment. These are light fabric funnels (usually made of rip-stop nylon) which direct the breeze into the cabin through any convenient hatch. The simplest, most effective type is like a small but lanky spinnaker, and must be adjusted to face the wind. A more sophisticated arrangement (Fig D35) is a four-sided scoop which works regardless of wind direction. Although a little more troublesome, the simpler type shifts the greatest volume of air. When at anchor in non-tidal waters, sailboats lie head-to-wind, anyway.

Once set facing forward, the single scoop type will continue to work without further attention.

Insect Screens

With the exception of man himself, mosquitoes are the most dangerous creatures on this planet. They kill more people than all the sharks, snakes, maddened bulls, and man-eating tigers put together. Less tragically, they are a plague upon paradise, guaranteed to spoil even the most idyllic anchorage with their evil little ways. And, unless you think they are exclusively a tropical menace, you ought to try the Baltic or Alaska on a tranquil summer's evening.

One way of dealing with them is to anoint oneself with a chemical repellent – typically, a disgusting concoction, the sensation of which is only barely less unpleasant than being bitten. Another answer is to burn mosquito coils – okay in an open cockpit on a calm night, but likely to make the cabin reek like a Hebridean kipper factory if used below. Or, if of a mechanical frame of mind, you might like to try your luck at keeping the little demons out with insect screens applied to all opening windows, hatches and vents. You are unlikely to be successful but they tell me that the therapeutic effect of 'doing something positive' is worthwhile. Personally, I have always found that squadrons of the varmints always managed to sneak in before I could secure the final seal.

Dining aboard an Australian yacht in the Balearics, I once sought to dip into that collective well of wisdom shared by cruising sailors. It was a velvet evening, as black as a coal-face, save for the glittering band of the milky way above. Anchored in Formentor, the water lay like glass between ourselves and the shore. The meal had been superb and we were now chatting in the cockpit, sinking the last of the Rioja and having a generally good time. And so were the mosquitoes, which dived like Stukas from the heavy pine-scented air.

'What the hell d'you do with these things?' I asked angrily, swatting at an insect now boring into my leg.

The Australian chuckled and my heart took flight. Bushmen were wise about such things, I felt sure. He turned to his wife and nodded. She went below and returned almost immediately with a square brown bottle which she handed to him.

'There you go, sport,' he said. 'Mozzie antidote. For internal use only.'

And a very fine brandy it was too.

Chapter 6
Self Steering

He was young, eager, and graduating upwards from dinghies. I was in need of a long weekend away and had recruited him as crew. Shortly after dawn, we cleared the buoyed channel and set course towards France. My young friend was at the tiller.

'You don't mind, do you?' he asked.

'Mind?'

'Me steering.'

'Absolutely not. Help yourself.'

'Oh, thanks.' His face glowed with gratitude as he murmured: 'Just for a few hours . . . or maybe more'.

And, I remembered being like that myself. For when I was new to sailing it seemed to me that steering was what it was all about; man and boat in perfect harmony, in intimate communion with wind and waves, and other such romantic maunderings. I still enjoy a trick at the helm, of course – at least for short spells – but as hour follows hour, the thrill wilts and then founders, leaving boredom in its place. To steer when it pleases you can be immensely satisfying, but to have to do it for extended periods is, in my view, unacceptable drudgery. Sadly, I thought, the lad now so blithely relishing his watch would learn that lesson soon enough. I was right. He was a quick learner. There was nothing like as much enthusiasm on the night passage home.

But, of course, this is more than a matter of simple pleasures. Being a slave to the helm imposes serious restrictions on your other responsibilities. Navigation, in the full sense of that word, is a complex business involving many disparate functions – most of which require the application of imagination and intelligence. Steering, on the other hand, is a fairly mindless task which can be very adequately performed by a machine.

Self steering systems are divided into two distinct groups: Windvane gears and electronic auto-pilots. Let's consider them in that order.

Self Steering

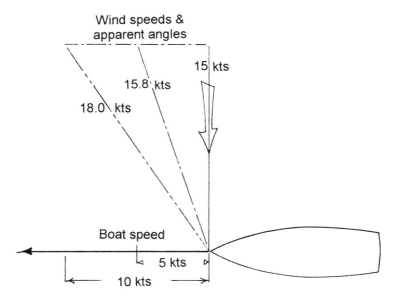

Fig D36 Apparent winds

Windvane Gears

These are entirely mechanical devices which sense the apparent wind direction and hold the vessel on a course relative to it. Windvane gears do not steer a compass course. If the wind direction changes, the yacht's heading will alter accordingly. Many the single-hander who has woken after a nap to find his boat on a wayward tack – some have even run hard aground. To guard against this, off-course alarms (more of this in Chapter 13) are often employed.

In order to work, windvane gears must perform two separate functions which can be described as 'sensing' and 'steering'. We shall deal with sensing first but, before we move on, we should cover that crucial little phrase 'apparent wind' – a phenomena which influences many aspects of sailing dynamics, not least windvane gear operation. As all sailors soon learn, the apparent wind is a combination of the true wind and the boat's movement through the water – 'apparent' as it would seem to someone standing on deck.

Figure D36 shows a yacht with a 15 knot true wind dead abeam, first making five knots and then ten (a light multihull, perhaps). The movement of the yacht brings the apparent wind forward of the beam, with an associated increase in velocity. And the faster the yacht sails, the greater will be this effect. In extreme cases, very fast sailboats can bring the apparent wind so finely onto the bow that they are effectively close-hauled, even though actually still sailing on a beam reach. It can easily be appreciated that the factors that determine the apparent wind are the yacht's course in relationship to the true wind, the velocity of the true wind, and the speed of the yacht. If any of these variables change, so will the apparent wind.

And, of course, the air-mass doesn't move along in conveniently straight lines. The atmosphere is a swirling gaseous soup, riddled with inconsistencies. The windvane gear sees only one thing – the apparent wind – and will strive to keep it at the same angle, regardless of where the boat might actually be going. Although at first this might seem unfortunate, remember that your sails are also trimmed to the apparent wind so, although your course over the ground might weave around a bit, you will at least be maintaining the optimum sailing angle. In fact, because they are almost instantly responsive to wind shifts, a well adjusted windvane gear will sail a boat to windward better than most helmsmen.

The simplest sensor is a plywood vane which pivots around a vertical axis, rather in the manner of a weathercock (Fig.D37). With the sailboat on course, the vane is allowed to feather into the apparent wind before the linkage is engaged. Once popular, this arrangement is now rarely used in commercial gears. Despite large vane areas and friction-reducing bearings, the vertical axis vane's rotational output is feeble in both power and degree. Particularly in light conditions (or running downwind with diminished apparent wind velocity) the yacht must yaw significantly off course before the deflection of the vane is sufficient to be useful.

Most modern windvanes now use a horizontal axis (actually, a slightly inclined axis) vane which is a considerably more sensitive and powerful sensor (Fig.D38). The vane assembly is mounted on a turn-table which is rotated to align it with the apparent wind – a slightly more troublesome task than simply allowing a vane to feather. A counterbalance holds the vane upright when so aligned, but the slightest deviation in course presents one face or other to the wind, which pushes it over. The leverage involved in this action can readily be imagined, and is delivered at even small angles of attack.

The wind direction data obtained from the vane must then be transmitted via a system of levers or lines to the steering part of the unit. The geometry involved in these linkages is too arcane to go into here. Sufficient to say that this is a critical part of the whole unit, involving differential rates of movement and what is known as 'negative feedback' to reduce oversteering. Do not be deceived by the apparent simplicity of these mechanisms; there is more to them than meets the eye. Some gears have adjustable linkages so that you can fine-tune to suit different conditions.

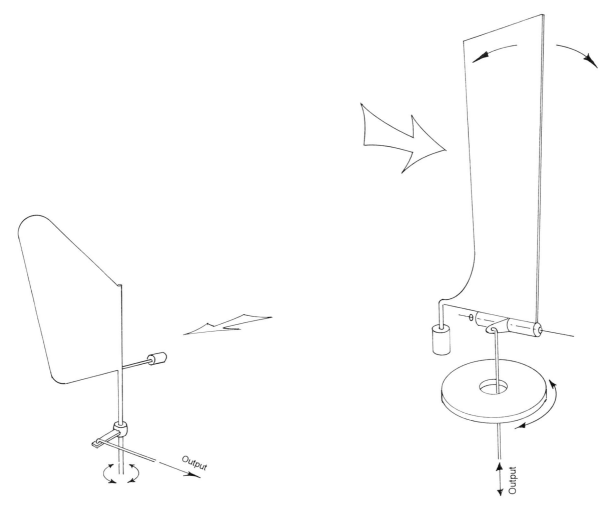

Fig D37 Vertical axis sensor

Fig D38 Horizontal axis sensor

Now we should turn to the bottom end of the windvane gear – the bits that steer the yacht. There are three different systems in general use, with a few hybrids which combine features from others. But the principal categories, and their advantages and disadvantages are:

Auxiliary Rudder

Here a generously sized vane (almost always of the more powerful horizontal axis variety) directly controls a balanced auxiliary rudder which, in turn, steers the yacht. There is no need for any connections to the main steering, which is usually lashed amidships or slightly to weather to ease the helm. The Hydrovane is a good example of this type.

The Innovative Yacht

Advantages

- The unit is neatly integrated and can be shipped or unshipped in a trice.
- There are no tiller lines to rig – a definite plus on a centre cockpit boat where these must be led forward.
- The absence of tiller lines also leaves a nice clear cockpit.
- The auxiliary rudder makes a useful reserve in case of main steering failure.
- No modifications to the yacht itself are required.

Disadvantages

- This is the least powerful type of windvane gear. As it must derive all of its operational power from the vane, it is clear that performance will suffer as the wind drops.

Trim-tab on Main Rudder

Perhaps the most practical type for sailboats with transom hung rudders. Figure D39 shows an arrangement utilising a vertical-axis vane and a very simple linkage. More sophisticated gears usually use horizontal-axis vanes but, of necessity, the linkages then become more complex.

As the vane is only required to move the relatively small tab, it need not be as huge as on the auxiliary rudder type. This is the most basic method of harnessing the water flow to do the donkey work. Essentially, the tab steers the rudder which then steers the boat. There are not many proprietary gears of this type on the market, but the Auto Steer by Clearway Design is a good example.

Advantages

- A robust and reliable system which can offer good performance at low cost. Trim-tab gears are often home built, and are usually easy to maintain and repair.

Disadvantages

- In the first instant, the action of the tab accentuates the yaw. Trim-tab gears can 'hunt' more than other types. Because of this they are probably more suitable for traditional long-keeled yachts, rather than the more skittish modern designs.
- Modifications to the rudder are usually required. The trim-tab gear is very much part of the boat, rather than just an add-on.
- Parts of the mechanism remain permanently underwater. A snorkel, mask, and scraper are useful additions to the maintenance kit.

Self Steering

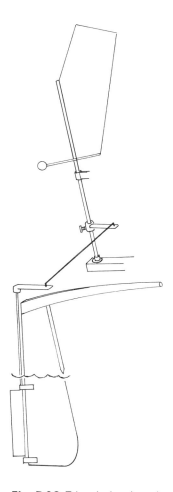

Fig D39 Trim-tab steering gear

A *Monitor* pendulum-servo gear. (Note also the solar panel.)

Pendulum-Servo Gears

These are the premier windvane gears, and will be the first choice for everyone interested in really top-notch performance. A narrow blade trails in the water astern. It is capable of turning like a rudder and also swivelling around a longitudinal axis (Fig.D40). When the boat strays off course, the vane (either vertical or horizontal axis) turns the pendulum which veers to one side in the water flow. By a variety of means, depending on individual design, lines from the pendulum are led to the tiller (or wheel) which operates the main rudder and returns the yacht to the straight and narrow.

The Innovative Yacht

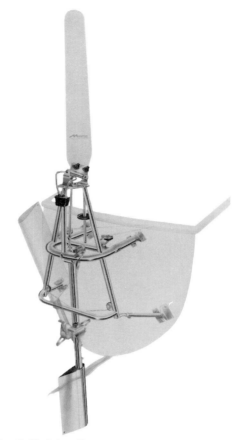

Fig D40 *Monitor* pendulum-servo gear

Pendulum-servo gears are immensely powerful bits of kit. I once nearly lost a finger in the works of one. The amplification of force between the relatively puny output from the vane and the mighty pull of the tiller lines, is hard to imagine.

Hardly surprisingly, most manufactured gears have been of this type. The most popular was undoubtedly the Aries – of which thousands are giving good service – but, alas, it has recently gone out of production. However, second-hand Aries are still to be found.

The pick of the current bunch is undoubtedly the Monitor™ – and a very fine pick it is too. Built in Sausalito, California, by Scanmar Marine, it is a beautifully designed and engineered piece of kit, of a quality which will appeal to all long distance sailors – many of whom have testified eloquently to its virtues. Made of stainless-steel, there is the minimum of intricate special parts. This is a tremendous asset for the cruising yachtsman who likes to know that, in the event of damage, the gear can probably be mended wherever he might be.

Advantages

- Great power and sensitivity. Pendulum-servo gears will perform across a wider wind velocity range than any other type. They are very responsive and will handle even the most twitchy boat if properly set up.
- All parts of the unit can be lifted clear of the water when not in use or for maintenance.
- Unlike the trim-tab type (which steers the 'wrong' way), the pendulum contributes to yaw correction by simple steering action as well as the servo function operating through the tiller lines.

Disadvantages

- Expense. Pendulum-servo gears are typically more complicated than other types.
- The pendulum is a slender blade, obviously vulnerable to damage, usually from heavy flotsam. Most have a weak point or shear-pin which will fail before the blade itself breaks or more serious damage is transmitted to the mechanism. Also, the inherent complexity of these gears increases the potential for mechanical problems.

As I mentioned earlier, although these are the main categories, there are also other variants. To name some examples: A double-servo gear with a small tab on the pendulum (the diminutive but powerful Navik); a servo-pendulum gear operating an auxiliary rudder (the Swedish Sailomat); and an auxiliary rudder with trim-tab system (Auto-helm). All use combinations of the basic principles as described above, and have similar attributes to the types from which they derive.

Most windvane gears can be adjusted remotely by control lines, which can even be led below. Of course, you will still need to go on deck to ship the vane or lower the pendulum blade, but the ability to make small course changes without going aft is very worthwhile.

At the end of the day, there is no such thing as the 'best' windvane gear – only the best for each particular boat. For instance, if the main steering is very stiff or slow, then a stand-alone auxiliary rudder gear may be the only answer. Certain gears can be adapted for wheel steering better than others. And, on ketches or yawls where the mizzen boom overhangs the stern, vane height will have to be considered.

Also, remember that there is no substitute for quality. Even the crudest windvane gear is quite a delicate piece of machinery, and it could be asked to steer for many thousands of miles without complaint. Your injection moulded plastic 'Wonder Wake Wizard' might look just the job in the mail order catalogue, but could leave you with a tiller in your hand when you would prefer to be cradling a gin and tonic.

Electronic Auto-pilots

If you can satisfy their appetite for electricity, the electronic auto-pilot can be a more convenient alternative to the windvane. For the short distance sailor, who routinely spends enough time under power to have electrical capacity to spare, it could be the first and most sensible choice. Even if ocean cruising, an auto-pilot can usefully complement a windvane, steering when motoring or motor-sailing. Many experienced yachtsmen carry both.

Auto-pilots come in two basic types. The first is known as a 'course follower' and the second a 'compass follower'. With the course-following type, the boat is manually brought round to the desired heading before a button is pressed to engage the auto-pilot. Thereafter it will maintain that course until a new one is required, whereupon the procedure must be repeated. With the compass following type, the course can be set directly on to the auto-pilot and changed at any time.

All auto-pilots use compasses as their primary directional sensors. In the early models these were usually of the conventional magnetised needle type, with movements being tracked by a photo-electric cell optically registering black and white sectors on a pivoted card. But modern units use what are known as fluxgate compasses. These employ an arrangement of fixed coils to sense the earth's magnetic field. Fluxgates have no moving parts (except for gimbals to keep them horizontal) and are extremely robust and reliable. On the very largest vessels, auto-pilots are controlled by gyro compasses – the only type which will relate to true (rather than magnetic) north.

In the smaller units, the fluxgate is usually built in. But, increasingly, auto-pilots are being designed to rely on directional data (possibly amongst other inputs) imported from a remote compass tucked away somewhere safe. This is part of the trend towards totally integrated electronic navigational systems – a subject on which we shall have more to say in a later chapter. Many auto-pilots also have the capability to accept wind direction information from a dedicated sensing vane or (again, integration here we come) from the yacht's centralised instrumentation. When operating in that mode, the fluxgate will be disarmed – or at least will impose only a secondary, monitoring influence on the course – and the yacht will be steered to the apparent wind in much the same way as a windvane might do the job.

Over the past few years, the sophistication of even the most basic auto-pilots have advanced considerably. The relatively crude course correction action has been enhanced by 'intelligent' software which will assess the prevailing conditions and match the operation of the auto-pilot to suit. And, all sorts of additional functions are now common. The list grows as manufacturers compete but, for a sample menu:

Remote Controls
Extra control units can be placed about the yacht. These are marvellous if you have a watchkeeping or steering position below. To be able to twitch a boat out of the

path of a threatening ship without first scrambling into the cockpit is wonderfully convenient.

Course Memory
And, having so twitched, it is also useful if the pilot remembers the previous heading and will return to it when required.

Response Control
This matches the sensitivity of the auto-pilot to the specific circumstances, usually by altering the 'deadband' limits – the off-course allowance permitted before the pilot takes corrective action. The least responsive setting obviously minimises battery drain and would normally be used unless navigational considerations dictate otherwise.

Compass Linearisation
This facility automatically smooths out errors caused by magnetic deviation. Deviation, mind – variation will remain, and your yacht will still be steering a magnetic course.

Watch Alarm
To help prevent you dozing off. A button must be pressed at least once within a prescribed period (typically every four minutes) otherwise a buzzer will sound and you will be cursed by the watch below.

Automatic Tacking
Pushing a button puts the boat about, leaving the crew free to tend the sheets. An obvious boon to the single-hander.

All proprietary auto-pilots operate through the yacht's main steering. Why this should always be the case beats me. A very small gear could work a tab on an auxiliary rudder (Fig.D41), with some very useful savings in electrical consumption. Perhaps someone will get around to developing such a system commercially but, in the meantime, it shouldn't be too difficult to devise.

For the moment, though, we must live with what's on offer. Today's equipment moves the boat's main rudder, and that takes power – ranging from quite modest for a small, well-balanced yacht, to voracious on a heavier vessel. The range of drive units offered by equipment manufacturers should cover all types of installation, and include: linear drives which work on a push-pull basis, operating either the tiller or a steering arm attached to the rudder stock; rotary units, providing a chain driven output – useful for torque tube type steering or when the rudder quadrant is too small or inaccessible for a steering arm; and, finally, hydraulic-drive units, which are controllable pumps plumbed into any hydraulic steering system. But, whatever

The Innovative Yacht

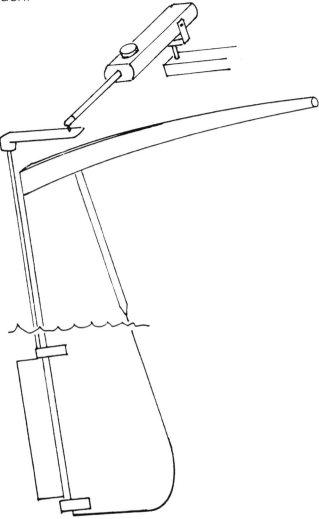

Fig D41 Auto pilot on a trim-tab

the method, the actual muscle is invariably provided by electric motors which (obviously) consume electricity.

And, here is the down side. Manufacturers' claims are not always to be trusted. Naturally, if a yacht is inherently directionally stable, and will need only the occasional tweak to keep her tamed, then electrical consumption will be low. But the average family yacht will have most auto-pilots sawing merrily away, sucking juice from the batteries like a child attacking a milk-shake.

Chapter 7
The Boat as a Habitat

I had to admit the visual effect was impressive. Removed from the hub-bub of Boat Show noises outside, the cathedral calm of the cabin exuded an impression of majesty and opulence. Along with other goggle-eyed supplicants, we had made an appointment the previous day to get aboard. Although not in the market for any kind of boat, let alone anything so splendid, we were impelled by that same vein of curiosity that brings people to view the homes of the noble and famous. Finely worked panelling shimmered in the indirect lighting. Rich Florentine patterned fabrics, quite worthy of a potentate's pleasure palace, set off the ceramic work surfaces and the swirling burr veneers of the saloon table. A handful of awed viewers padded around, voices lowered in reverential whispers, fingering the furnishings and brightwork, like pilgrims arrived at a shrine. We were in the presence of great luxury and the salesmen, blazered, duck trousered and smelling strongly of after-shave, all knew it.

My companion, a work-worn yachting journalist who claims to have sacrificed both humour and liver in pursuit of his profession, wandered about radiating the kind of distaste that Jonah must have displayed when newly swallowed. Grumpily, he poked about before approaching the sleekest of the sales team.

'Tell me, young man, where would I lash the dead sheep?' he asked loudly.

The salesman blinked. All movement ceased everywhere as his words ricocheted around the cabin. An attentive silence settled over the yacht.

'Pardon?'

Ignoring the 'No Smoking' signs, my friend took out his pipe and started to load it with a foul substance having the appearance of well rotted manure. 'Story time', he announced. 'When I was a nipper in the Merchant Navy just after the war, we took out this brand new, special order Rolls-Royce to Australia. It was a magnificent machine for sure. Gleaming coach-built body, walnut trim, and snow-white hide upholstery. Amongst its other wonders it also had air conditioning, a cocktail cabinet, and a little gadget that spat out ice cubes – just the sort of thing you'd want if you

The Innovative Yacht

owned several hundred square miles of prime grazing and felt like treating yourself. Are you following this?'

'Well . . . er . .'

'Anyway, when we got to Sydney and off-loaded this transport of delight, our drover chappy was there to meet it. What d'you think of that, eh?'

'I'm sorry, sir, but . . .' The salesman's eyes signalled for reinforcements. There were no takers.

'You'd guess he'd be tickled pink, wouldn't you? Just like the lucky sods that buy your boats. Well, you'd be quite wrong; the poor bloke was stricken. You see, his old Roller had running boards, to which he would lash any dead sheep he found. And the new car was of the modern style with flush topsides. For all the fancy odds and sods it had aboard, it was seriously deficient in his eyes. Now d'you get it?'

We all waited for a response but there was none coming. So hypnotised was the salesman by this manic encounter that he ignored the flare of the match and the cloud of smoke that filled that hallowed cocoon.

'All right, let me make it easier for you. How could I work at that chart table if soaking wet? Where's the sea berth where I can doss down if ditto? What would keep you in that silly armchair on starboard tack? Where would I stow the dinghy if I didn't want it in the davits? And, if you'll forgive the plain speaking, in sailing terms there's nowhere on this thing you could lash a dead sheep. This isn't a boat, laddy, it's a bloody floating boudoir!'

For the arrival of the interior designer into the world of yachts has not always been a cause for celebration. This is not to say that the accommodation is unworthy of the same sort of professional attention that should be lavished on the hull and rig, but sometimes the chintzier excesses of creative zeal can ruin an otherwise decent boat. Of course, flashy furnishing and an ingeniously commodious lay-out are known to be powerful selling points. The manufacturer, with his enormous capital investment in a new model, is under extreme pressure to compete by offering the most for the least – and by that he usually means the most berths for the least length and money.

To plot the development of this expedient corruption, let's take a fictitious but representative modern design – a 36ft (11.0m) sloop of generous beam and lightish displacement. The naval architect has done his stuff and handed the hull lines and sail plan over to his client, the builder. He passes them on to the interior designer with instructions to conjure up an accommodation which will satisfy the following criteria:

1. Sleeping for eight adults.

2. Two enclosed heads with showers, one of which must be en suite with the 'owner's cabin'.

3. A seating and dining area where the crew of eight might convivially congregate. Intended as the area where social events will focus, a visually striking approach was deemed desirable.

4. A galley of sufficient capacity to feed the aforementioned crew of eight. Equipment must include a double sink, hot water supply and refrigerator.

5. A navigation area with seat, full-size chart table, and an instrument panel vaguely reminiscent of Concorde's.

Well, mindful of his mortgage and the need to feed the kids, our man wanders off, switches on his computer and sets to work. Some weeks later he emerges with the results. Drawings, colour charts and swatches of fabric are laid out for consideration. Sales and production men gather around like archeologists about to unwrap a mummy. The tension is electric. But at last, just when our designer thinks he can bear the suspense no longer, the first smile breaks like dawn in the east and a flurry of enthusiastic back-slapping ensues.

And, it has to be admitted that his creation is a masterpiece of ingenuity. Every design objective has been achieved – actually, even exceeded, for there is an extra heads. As can be seen in Figure D42, not a square inch has been wasted. The marketing ideal of 100 per cent space utilisation has been well satisfied. In due course a wide-angle lens will produce wondrous glossies for the brochure. Magazine articles will praise the level of comfort. Future boat shows will see admiring hordes traipsing through, hopefully with their cheque books primed and ready.

But, if we can tear ourselves away from the world of sales-speak, and imagine that same boat afloat and cruising, how does the interior design stand up? In my opinion, not at all well. Let's deal with the features one by one.

- Eight berths is ridiculous for a vessel of this size and is, anyway, a rather dishonest description because we are actually not talking about eight individuals but four couples. The ideal crew would be four and the boat could easily (and is more likely to) be handled by two. And, as all the berths are doubles, where would you sleep on passage? I suppose it would be possible to split the cushions and divide the doubles into singles by using lee cloths, but this is an awkward solution. A sensible sea boat should have secure single berths for the watch below.

- The sanitary arrangements are also extravagant. Unless planning a mass sit-in, the loos-per-head factor is more than could ever be needed. Also remember that each toilet requires two seacocks, plus diverter valves to holding tanks if these are fitted. With all the various inlets and discharges taken into account, the total number of holes below the waterline (each fitted with seacocks which should be routinely opened and closed), could be as many as fourteen on this boat! Handling the plumbing alone would be a major responsibility.

- The showers are a nice idea but, again, impractical. The spray within those tiny cubicles would thoroughly drench the interior. This would necessitate putting everything away – toilet paper, towels, cosmetics, whatever – before use.

- That chic dinette area looks great on the drawing but would be very uncomfortable with the yacht heeled on either tack. Curved seating is hardly ever successful on sailboats. There are no corners in which to wedge oneself and that most comfortable of off-watch postures, sitting lengthways with your feet up, is denied you. On port tack, the settee would be uphill to windward and you might need to wear seat belts to stay there.

- The cook will not be thanking the designer in hot weather or if there is any kind of sea running. The galley is too far forward, where pitching motion is greatest, and away from the natural ventilation of the companionway. Cooking at sea calls for an iron stomach and involves juggling with boiling liquids. No opportunity to make the cook's life easier should be missed.

- The two aft cabins rule out all but the most paltry cockpit lockers. Indeed, the whole matter of stowage has been neglected on this boat. Where would you put sail bags, buckets, bicycles, boarding ladders, dinghies – let alone dead sheep? The modern light-displacement hull has very little volume beneath bunks, and there is an obvious limit to how much you can pile on deck. The likely outcome would be that one of the cabins would be adopted as a bosun's locker, so why not design it that way in the first place?

The flaws in this design arose from commercial imperatives. Boatbuilders live in a precarious trading environment in which the failure rate is high. You can hardly blame them for supplying the public with what they think they want as long as we continue to prove to them that they are right. The pity is that many of us continue to be seduced by the delusion that a quart really can be fitted into a pint pot.

Figure D43 shows a more seamanlike approach to the same project. One of the enclosed cabins and a brace of heads have been sacrificed in favour of more useful facilities. The seating arrangement works on either tack and can be lounged upon in proper languorous manner. The galley has reassumed its traditional position under the companionway and, abaft that, is an enclosed heads compartment of sufficient size to rig a more workable shower. Perhaps more importantly, there is now a huge cockpit locker capable of swallowing a mountain of gear. Sad to say, if both options were offered it is probable that this would be the least popular choice.

The Boat as a Habitat

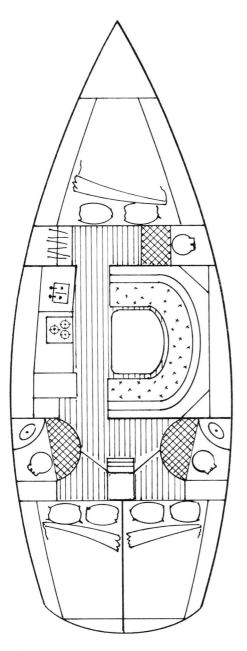

Fig D42 'Designer' interior layout

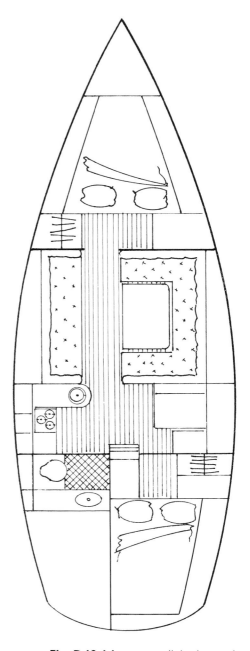

Fig D43 More sensible layout

The Innovative Yacht

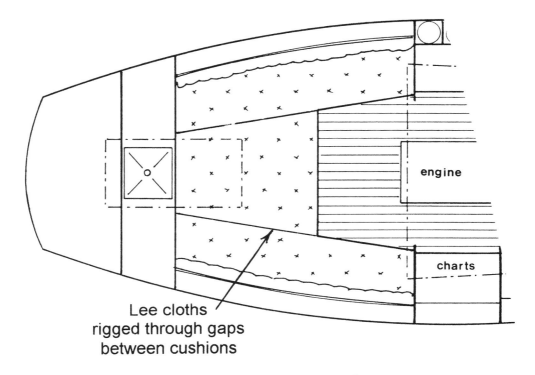

Lee cloths rigged through gaps between cushions

Fig D44 *Spook's* double berth

Of Bunks and Things

The cruising sailboat is both home and vehicle and should be designed to function well in both roles. One moment you will be at sea, your routine broken, your world slanted and heaving, and the next you will be alongside relaxed and comfortable, socially eager to spread yourself about a bit. This duality of function presents the designer with one of his biggest headaches, and nowhere is this headache more acute than in the sleeping arrangements.

It is one of the fundamental rules of life that double bunks are good things – even if you intend to sleep alone, and especially in hot climates where any restriction on air circulation is to be avoided. Personally, I prefer nothing better than an afternoon siesta spent extravagantly spreadeagled like a prematurely landed sky-diver. My wife claims that this is not a pretty sight, but seen from the inside it is heaven.

But, whilst delightful in port or in very calm weather, double bunks are pretty useless resting places at sea. Even if wedged against the hull side to leeward, the feeling of insecurity they inspire can keep many people awake no matter how exhausted they might be.

Our own boat, *Spook*, has a generously proportioned double berth – which also forms part of the seating area – running across the aft end of the cabin (Fig.D44). Obviously an athwartships berth would be even less tenable when heeled than a fore-and-aft one, so we rarely use it at sea. However, the cushions are divided in such a way as to allow longitudinal lee cloths to be rigged. In combination with the cabin seating, this provides a pair of very snug sea berths for passage making. In practice, we usually cruise two-handed, so only a single sea berth need be rigged – usually the windward one – for the person off watch. This leaves the more secure leeward side for the watchkeeper to sit between trips on deck to keep a look-out.

This principle can be adapted to most double berths. Often on production boats the cushions are divided laterally, which is not only less comfortable but prevents the berth being split into singles.

The Dedicated Sea Berth

I once sailed briefly with a man who seemed to enjoy being always soaked to the skin. Admittedly, it was hot, and presumably he found it cooling. But he made a mess of my boat and that made me mad. Down the companionway he would come, squelching through the cabin, and then flopping into whichever berth took his fancy. Within a couple of days they were all wet. I put him ashore in Tampa, Florida. We didn't part friends.

Here my higher latitude upbringing shows its colours. I am known to be obsessive about keeping the cabin dry. In my view nothing drains the morale like a sopping interior. And, in cold climates, it can be downright dangerous.

But, of course, the sea is a very wet place and it tends to get spread about. And we have all been damper than we would like to be – and in desperate need of a rest. The answer, if it can be devised, is a proper sea berth.

Figure D45 shows a very workable arrangement. The berth is located beneath a stand-up chart table so lee cloths are only necessary in the vilest conditions. Being almost directly under the companionway it can be reached without tramping through the rest of the accommodation. And the cushion is covered with waterproof PVC to minimise water absorption. Even a person in wet oilskins could rest there without creating misery for everyone else.

Crew Size

This is a pivotal matter when considering interior comfort. When pursuing the preliminary research for this book, I asked a friend of mine, who regularly sailed with his wife and four teenage children, what he would add to his boat to make life more comfortable. Without hesitation he replied: 'Another ten feet'.

Unless wishing to emulate a Ganges ferry, there has to be a relationship between the size of the yacht and the number of bodies that can reasonably be put aboard.

The Innovative Yacht

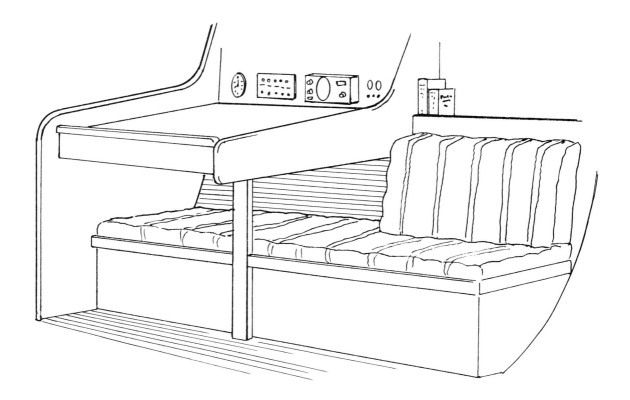

Fig D45 Dedicated sea berth under the chart table

At the risk of repeating myself, having eight berths in a boat does not necessarily make it an eight berth boat – it could be (and probably is) just a boat with too many berths. Personal preferences and tolerance thresholds vary but most experienced sailors are agreed that, commensurate with a yacht's handling demands, the fewer the merrier.

The second column of the table below was distiled from manufacturers' brochures, the third is a consensus of ideal crew size drawn from some of my saner and more distantly travelled chums, and the fourth is the just tolerable short term maximum as expressed by that same motley bunch. Discussion on those last figures produced the most conversational heat, with one man observing that he didn't actually think there existed seven other people he would want to go sailing with.

The Boat as a Habitat

Length Overall	Berths Offered	Ideal	Maximum
25ft (7.62m)	4–6	2	4
30ft (9.14m)	5–8	2	4
35ft (10.67m)	6–8	4	6
40ft (12.19m)	6–9	4	6
45ft (13.72m)	7–10	4	6
50ft (15.24m)	8–12	6	8

As can be seen, there is a disparity between what the designers and builders believe we need and the social realities. Sailing is as much about escape as it is about discovery. The more jarring problems associated with human relationships are often heightened by time at sea, and to find yourself trapped aboard an unhappy ship is to discover that you haven't escaped at all. It is up to each of us to make our own decisions in these matters but, like the guest room upstairs at home, tranquillity is often gained by leaving it uninhabited.

It is often of benefit to keep the galley and navigational area adjacent, both at sea and in port.

Chapter 8
Feeding the Five Thousand

'Unhappy is the crew who ignores the cook', a wise old friend once said; and his waistline certainly testified to his wisdom on the subject. It is the toughest, most thankless job aboard, and anything that can be done to lighten the load will be bounteously rewarded.

For those who are unfamiliar with the process, cooking involves juggling with very hot substances, and peering down into mixtures that could sicken Macbeth's witches. It is not work for the faint-hearted.

The principal requirements for an efficient galley are:

- The location must be well ventilated and accessible to both the seating area and cockpit. Directly beneath the companionway hatch is often ideal. Centre cockpit sailboats with a passage linking main and aft cabins very often place the galley in the passageway where the space is confined and the ventilation poor. This is usually because the passage is useless for any other purpose and, for marketing purposes, therefore considered a waste – an ungenerous motive for such a design decision.

- The galley should be positioned towards the stern where the yacht's pitching is felt least. There is nothing you can do about roll but, as the cooker should be gimballed on a longitudinal axis, this is largely taken care of. Again, the site beneath the companionway is usually good. In most layouts, this allows food to be readily handed to either the cockpit or seating area.

- There should be plenty of work space, including an area where you can set down hot pans. The whole working surface must be generally heat resistant, but one specific area should be especially so. Many people use the sink, but this seldom works well for pots with handles. A raised metal grid, or a recessed tiled insert would be better if it could be contrived.

Feeding the Five Thousand

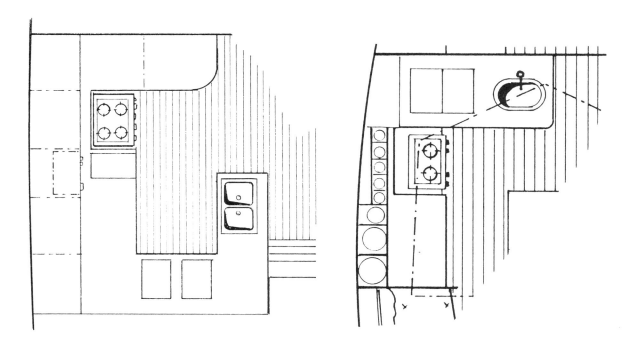

Fig D46.1 A good working galley **Fig D46.2** — and another for a smaller boat

- And the work surface – most particularly the 'hot zone' – should be fitted with generously proportioned fiddles. Those fitted to many production boats are woefully inadequate. 75mm (3in) should be the minimum.

- The cook (often a lady) must be able to either wedge herself into the actual structure of the galley or should be supported with a strap. She must be able to work with both hands at once – impossible if she also has to cling on. But beware! Thus secured she has no chance of stepping back if she spills something hot. And some very serious scalds have occurred this way. The rule on *Spook* is that if it is rough enough to warrant the galley strap, the cook wears the bottom half of her oilskins.

Figures D46.1 & 2 show two equally workable layouts for a galley – one more appropriate to a larger yacht than the other. But the characteristic they both share is that all of the vital elements of cooking – ingredients, utensils, crockery, cooker, sink and refrigerator – can be reached without moving more than a couple of steps.

Good ventilation is absolutely essential. If it cannot be achieved naturally through hatches or opening scuttles, then electric extractors should be considered. Remember that cooking produces large quantities of steam which will rapidly raise the humidity past acceptable levels and, once condensed, will dampen the interior and provide a perfect environment for mildew. And that not all cooking aromas are pleasant – especially if they linger well past any nostalgia for the meal.

Cookers

These are quite obviously the heart of every galley and should be chosen with at least as much care and discernment as you might apply to your sails, winches, or any other vital part of your yacht. A well designed, sensibly built cooker will not only be a pleasure to cook on, but will go a long way towards protecting you against the very real perils of cooking at sea.

The various fuel types with their virtues and vices can be summarised as follows:

Liquid (or Liquefied) Petroleum Gas (LPG)

The two types in widespread use are propane and butane, and the choice between these will depend to some extent upon local availability – some places favour propane, others butane. And some proprietary LPGs are a mixture of both. The more remote the location, the less choice you will have, and the more you will be at the mercy of local suppliers. If long-distance cruising, it pays to look ahead. Many sailboats carry a selection of different connectors so that they can hook-up to a variety of manufacturers' cylinders.

The principal difference between propane and butane is that the former will evaporate within the cylinder into a usable, combustible gas at a lower temperature than butane. In extremely cold conditions butane may fail to work at all. But, for more average climates, either will do and they can be considered interchangeable.

LPG is probably the most commonly used cooking fuel for Britain and much of Europe, but it is viewed with vast suspicion in many other parts of the world. And the risks certainly need to be respected. When mixed with oxygen in the air, LPG forms a dangerously explosive mixture which can literally blow a boat to bits if accidentally ignited. It is much heavier than air, and can accumulate in the bilge until it reaches dangerous concentrations. All it would then take is one small spark – perhaps just from static electricity – to wreak the most awful damage to boat and crew.

Many accidents happen when, unbeknownst to the cook, the flame is accidentally extinguished and gas continues to pour from the burner. A pot may boil over or be spilt by the motion, or a gust of wind may blow down the companionway, snuffing out the flame. Either way, it is obviously bad news for the safety of the vessel. The better modern cookers are fitted with flame failure devices (thermocouples) which will

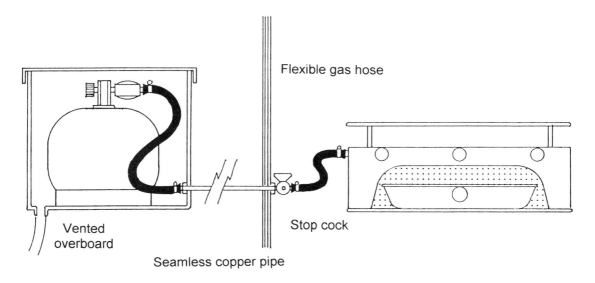

Fig D47 Gas installation

only admit gas to the burner so long as a minimum temperature is maintained. If the flame goes out, the burner cools rapidly and, within seconds, the gas is shut off. Some manufacturers offer a range of cookers on which the cheapest might have no flame failure devices, the mid-range models perhaps only one in the oven, and the top of the line with all burners protected. In my opinion, it is madness to settle for less than the very best. And I am sure it won't be long before insurance companies insist on it!

Figure D47 shows a typical installation. Ideally, this should be backed up with a gas detector ('sniffer') with sensors in the bilge. LPG is only as dangerous as we allow it to be. If properly installed and regularly maintained, it is a superb fuel. It is both efficient and convenient. It has a very high heat output (approximately 2,100 Btus per lb.), is odourless if fully combusted, and can be turned on and off as handily as your domestic cooker at home.

Compressed Natural Gas (CNG)

This type of cooking gas is virtually unknown in Europe but enjoys more popularity in the United States. Whereas LPG liquefies at relatively low cylinder pressures, CNG remains gaseous and, as the name implies, is simply compressed into its cylinder in much the same way as a SCUBA tank is filled.

Promoters of CNG will tell you that, as the gas is lighter than air, any accidental spillage will rise and disperse, thereby making it a safer option to LPG. A more cynical observer once said that it simply means you get your head blown off rather than your feet.

In terms of heat output per unit weight, CNG is less than half as efficient as LPG, and is nothing like as readily available world-wide. For the long-distance sailor, it has little to commend it.

Paraffin Oil or Kerosene

After LPG, this is perhaps the next most popular cooking fuel, and one for which I have an almost perverse affection. Messy, smelly, and in some places (Spain, for instance) astonishingly difficult to buy, it is nonetheless a very useful fuel which cooks about as quickly as LPG. It is also very safe.

Paraffin cookers invariably use a pressurised tank that feeds the burners via a route that passes the fuel through the heat of its own flame – thereby evaporating it into a combustible vapour. In order to get this self-perpetuating process rolling, the burners must first be pre-heated with alcohol (methylated spirits) or some other means – small gas cartridge type blowtorches are very effective. Paraffin cookers tend to be temperamental – which is perhaps why they intrigue me – and need coaxing and cajoling as might a wilful child. On a more practical level, they also need regular maintenance to keep the jets and other bits clean and clear. But, get a good one and it will serve you faithfully for years.

Alcohol (Methylated Spirits)

A rather wimpish fuel but with some advantages. Alcohol comes in various forms, and it is a bitter irony that the magic ingredient of booze (ethyl alcohol) is also the most suitable as a fuel. Now, for reasons of taxation and health (probably in that order) our lords and masters in government don't want gallons of the stuff available at low cost – ostensibly for heating the porridge but also potentially an ingredient for all manner of cheap cocktails – so unsportingly adulterate (or 'denature', as it is called in the US) the ethyl alcohol with methyl alcohol to make it undrinkable by all but the most determined.

Of all fuels, alcohol is generally the safest. Unpressurised, it cannot explode, and its flame lacks the ferocity of LPG or paraffin oil. The flame is also a delicate thing which can be snuffed out by anything vigorous in the way of draughts. But its main deficiency is its lack of heat output. It burns relatively cool and consequently takes longer to cook anything – hardly the characteristic you would want of a serious fuel. And, of all the fuels, it is usually the most expensive.

But, for people who only cook occasionally, alcohol could be a good choice.

Alcohol/Electric

A hybrid arrangement which could be of interest to those who spend a lot of time hooked-up to shore power. Alcohol burners are located directly beneath 110v electric elements (Europeans would need step-down transformers). When alongside, the cooker is used in exactly the same way as you would at home. Away from the marina the alcohol comes into its own.

So far as the cookers themselves are concerned, it pays to go for the best quality. The features to look for are:

- Solid, corrosion-proof construction. Your camping-type stove, with its pressed-steel shell simply won't do.

- Adequate gimballing to allow for angles of heel. This is largely up to the boatbuilder, of course. On some yachts the cooker has virtually unlimited swing on one tack but crashes against the outboard face of its recess on the other. And, if an LPG or CNG cooker, the supply hose must be protected from damage as the cooker swings. If it pinches in any way, then it is only a matter of time before the hose is breached.

    And it is also important that the gimbals can be locked upright for times when the yacht is in harbour. The very feature that adds to safety when at sea, can be a downright menace alongside.

- Sturdy fiddles and/or pan clamps must be fitted. Those seen on too many cookers seem to have been added as an afterthought, with no real understanding of what will be expected of them in service. A low rail around the entire hob area is not enough. Imagine preparing spaghetti for six in your largest pan, and then think of it all sliding about. Anyone for salad?

- The grill (or 'broiler' in the US) pan must lock positively under its burners when in use. The flat ledge type with nary a lip or latch really is a nonsense. Add to that a protruding handle which any passing crew member could nudge and you have the makings of a highly mobile projectile.

- On any gas cooker all burners should be fitted with flame failure devices. I implore you to look elsewhere for cost-savings. As a marine surveyor, I have probably seen more than my share of burned out wrecks, and I can tell you that a gas explosion is not something you would want to experience first hand.

You really do get what you pay for in the cooker line. There is a discernible correlation between price and quality. The caravan type cooker might be fine in your camper van but has no place at all on your sailboat. Unfortunately, boatbuilders are under great pressure to keep costs low, and they know that one cooker can look very like another superficially. When a top quality cooker might cost about £1,000, and they can buy another unit that will do the same job for a third of that sum, the savings are very tempting indeed. And the upshot of this is that many production sailboats hit the market, bristling with fine equipment in all other departments, whilst the cooker is a sorry piece of gear in no way matching the rest.

The Innovative Yacht

Our charcoal burning barbecue gets frequent use on *Spook*. This particular model is manufactured by Force 10 Marine Ltd, of Canada.

Microwaves

Really an alongside device but still useful on the larger yacht. Although some microwave ovens will operate on 12VDC, they are usually domestic-type units that have been converted by building-in invertors to raise the boat's voltage to that of shore power. At sea, many people would find the electrical drain unacceptable, but big boats may well have a 110VAC or 240VAC generator aboard which could be powered up when its time to cook.

Barbecues

But cooking remains a grisly business, and never more so than in the heat of a tropical day when 'sweating over a hot stove' becomes more than just a cliché. And it's not just the cook who suffers. Even a simple meal cooked below can raise the cabin temperature to unbearable levels. When cruising the Caribbean there were times when we were forced to bake our bread at night, simply to keep ourselves from melting.

The shipboard barbecue is a marvellous bit of kit – even in temperate latitudes where chilliness is more usually the problem. To have the day's catch straight off the hook and onto the coals, is to really know what fresh fish tastes like.

Some yachts carry those foil packed disposable barbecues but, personally, I find them unsatisfactory. There is rarely a safe enough space to fire them up on a boat and, anyway, the chemical used to make them self-igniting taints the food unless it is left a considerable time to burn away.

The properly constructed stainless-steel yacht barbecue is a far more useful device.

Pressure Cookers

The ocean voyaging cook's favourite utensil. They have been around for years, but should still be strongly recommended to all that have sailed without one. My wife, Chele, is addicted to globe artichokes, and suffers excitement almost embarrassing to behold when she sees them so profusely and freshly marketed in French ports. Conventionally cooked, they take upwards of forty-five minutes and require ridiculous quantities of gas and water. But popped into a pressure cooker, they take only a fraction of that time, consuming no more than a cupful or two of our precious drinking water. Obviously, similar economies can be made with many other meals.

Used as a conventional saucepan in rough weather (with the pressure valve removed), the added security of the locking lid makes cooking very much safer.

Refrigeration

I have always found the gentle clink of ice against glass to be one of life's most cheering sounds, capable of enlivening even the dreariest anchorage. In any other guise, ice would be eschewed by all self-respecting sailors. To have it encrust the topsides or float past in city-sized lumps is definitely to be avoided. But a plentiful supply, employed to lower the temperature of the local tipple and keep perishable foods fresh, deserves serious applause. Anyway, I am married to a Texan and, if I am to remain that way, some means of keeping things cool is obligatory.

Of course, you can import the ice from ashore and store it in an insulated compartment, but this has obvious limitations. In the Caribbean we found we could make a large block last for four or five days if we disciplined ourselves not to open the icebox too frequently. But replenishment was always a problem and, to our shame, we found we were scheduling our itinerary in a series of hops between ports where we knew it to be available – and that was not as easy as it sounds.

But, if you want cold storage and independence, you need onboard refrigeration. For the boat owner, there are three possible approaches. The first two are closely related so we will deal with these first.

Compressor Systems – How They Work

A sunny day and T-shirt sailing. The air temperature is comfortably warm and so are you. Then suddenly, along comes a shower and it rains heavily for a few minutes before the sunshine resumes. At first you are refreshed, but then you feel chilly and soon your teeth are chattering. Why?

The answer lies in the 'latent heat of vaporisation'. As the water in your shirt evaporates, large amounts of energy (in the form of heat) are required to convert it from a liquid to a vapour of the same temperature. This heat is supplied partially by the sun but also by your body which consequently gives up that heat and is chilled. It is, incidentally, the way nature controls excessive body heat by sweating.

Taking another model: as any mountaineer will tell you, it is difficult to make a good cup of tea on a mountain-top. This is because in the reduced atmospheric pressures found at altitude, water boils at too low a temperature. And, no matter how long you leave the kettle on the stove, the temperature will not rise above that lower boiling point. All the energy will have been utilised in converting the liquid to steam.

These two concepts – latent heat and the changes in the boiling temperature with pressure – are central to the operation of most refrigeration (and air-conditioning) systems.

Figure D48 shows the main components schematically. It is essentially an enclosed circuit around which a liquid is circulated by a compressor. Various liquids have been used in the past (ammonia being a rather unpleasant example) but most units these days use a refrigerant known as Freon R12, a trade name of the DuPont company. However, the link between fluorocarbons and the depletion of the ozone layer has prompted moves to encourage the use of less anti-social substances. The characteristics that all modern refrigerants share is that they boil at low temperatures (liquid R12 at –29.7°C (–21.6°F) at atmospheric pressure).

Let's look at the sequence of events in stages:

1. The compressor draws in vaporised refrigerant gas and pumps it on to the condenser.

2. The condenser, which is cooled either by air or sea-water, cools the gas and it condenses into a liquid. In so doing it gives up latent heat of condensation, which is carried away by the condenser coolant (the air or sea-water).

3. Next in line is the expansion valve. At its simplest, this could be nothing more than a piece of capillary tubing with a very small orifice. Other, more elaborate, devices have orifices which are adjusted automatically, controlled by a heat-sensing bulb thermostat to accord with the system's needs. The refrigerant is forced through the valve, emerging into the evaporator cold plate in a fine spray.

4. It can be seen from the diagram that the evaporator is connected to the low pressure, suction side of the circuit. The sudden drop in pressure between the upstream side of the expansion valve and the evaporator instantly causes the liquid to boil into a vapour (remember the mountaineer's kettle). In doing so it absorbs large amounts of latent heat from its surroundings which, of course, is the inside of the refrigerator and its contents.

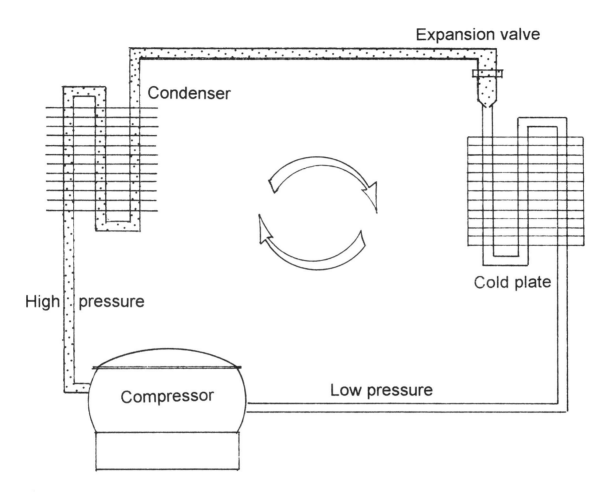

Fig D48 Schematic refrigeration

5. The heat-bearing gas returns to the compressor and goes on to the condenser where the heat is extracted and dissipated. And so the cycle is repeated.

Now that we understand the way the system functions as a whole, we can consider the individual components and their variants, some of which are more suitable for certain applications than others.

Compressors

These are either electrically powered or driven by a belt from the main engine.

Electric compressors come in two types – piston and diaphragm. A piston compressor resembles a small reciprocating engine and, like the engine, can be stripped and serviced by a competent mechanic. The diaphragm compressor is less costly to manufacture but is almost invariably a sealed unit which must be replaced when it fails. Because of its compactness, efficiency and quiet operation, the latter type is by far the most popular for all but the largest installations. Nearly all proprietary 'packaged' systems use this option.

On sailboats with limited electrical resources, the drawback of either type of compressor is the draw on the batteries – which can be savage if not properly regulated. Later in this chapter we shall cover controllers which assist the skipper in getting the best out his generating and battery capacity but, for the moment, let's just agree that in electrical terms, refrigerators are exceptionally greedy.

The engine driven compressor is usually of the swash plate type, developed from automobile air-conditioning units. These are engaged through an electromagnetic clutch, actuated either by an automatic control system or by simply throwing a switch. Engine compressors are very powerful and obviously make no significant demands upon the ship's electrics (the thermostat circuits will always draw a little). However, they are entirely dependent upon running the engine – perhaps not always convenient in a sailing yacht, especially when lying alongside in the marina.

Condensers

The variant here is the cooling medium – air or water. Of these, water is by far the better conductor of heat, and therefore the most effective.

The simplest (and least costly) units pass the refrigerant through vaned coils – either passive or wafted by a small fan (similar to a car's radiator). More efficient condensers are enclosed heat exchangers which draw sea-water as a coolant from outside the boat.

In hot climates, the air-cooled condensers may run into problems. The difference between the condenser temperature and the ambient air temperature might not be enough to conduct the heat away. Sea-water would have no such difficulty.

Cold Plates

The choice here is between the basic evaporator plate and the more substantial cold storage or 'eutectic' plate.

Evaporator plates cool down very quickly, but are equally quick to warm again. Such a refrigerator would need to be responsive to a thermostat which would cut the compressor in and out as the temperature rises and falls. This is perhaps the best type to make those ice cubes we dreamed about earlier. But expect the compressor to be busy and current drain high.

The eutectic plate has the evaporated refrigerant passing through coils within a sealed – usually stainless-steel – container filled with eutectic liquid. Eutectic, incidentally, means 'freezing and melting at minimum temperatures'. In operation the approach is different from the previous type. Here the temperature of the plate is pulled down in one hit – usually with an oversized compressor which, ironically, is more efficient in terms of cooling power gained for the amount of electricity consumed. It then switches itself off and the cold plate slowly releases the cold to the space within the refrigerator until the temperature rises to a point where the cycle is initiated again.

Eutectic plate fridges are ideal for sailboat use but the minute-by-minute control over the temperature is impossible. The cold storage plate is either freezing, frozen solid, or slowly thawing. This makes the design of the fridge fairly critical, with the plate sized-matched to volume, and the freezer and cooler compartments carefully arranged with regard to distances from the plate. Even so, there can be an operational lack of flexibility in these units. An installation I designed for use in the Mediterranean was so aggressively efficient in the temperate English spring that we had cucumbers that rang like bells!

Control and Battery Protection

So far as electricity is concerned, sailboats are inconsistent things. With the engine thumping away, you have bags of it to spare. Then, a little later, you are under sail with your autopilot and instruments on and your batteries in decline. Unless you have some other means of charging them (Chapter 12), this depletion will end in total discharge, no matter how stingily you eke them out. Sailors are understandably wary of such an outcome.

And, this in turn comes as no surprise to the manufacturers of marine refrigeration, who know that they are competing not just for the customers' cash, but also their perception of how much electricity they can afford to squander. As sailboats become ever more reliant upon electronics, the need to evaluate priorities becomes more pressing.

However, in one important respect, fridges have it over other equipment. Switch something like an autopilot off and it immediately ceases to be useful – there is no residual function living on to steer the boat. But the icebox – especially one fitted with an eutectic plate – will continue to keep the beer cool for many hours to come. Therefore, if you can cool the unit down rapidly when the engine is running and there is a surplus of electricity, the cooling effect could freewheel through until the next time you fire her up.

Your 'intelligent' controller arranges just that. A typical program could be configured thus: When the engine is running, the alternator will raise the charge level to about 14 volts, and the controller will sense this and switch on the compressor. Typically, it might take about an hour to get a 150 litre (5.3 cu.ft.) fridge down to –15°C (5° F). Whilst the engine thunders on, the thermostat will hold the temperature between –8°C (17.6°F) and –15°C. But, if the engine is shut down before this is achieved, the controller will settle for a less ambitious cool-down level of between −1°C and −6°C (30.2°F and 21.2°F). Thereafter, the compressor will give the icebox a boost of about five minutes per hour to help maintain the cooling plate temperature. In any event, the unit will switch off completely to protect the batteries if their charge levels fall below 10.5 Volts.

Engine driven compressor controllers work slightly differently, for obviously there is no top-up possible when the engine is still. Taking advantage of the compressor's extra power, the aim here is to pull the eutectic plate down to −15°C and hold it there until the very last moment.

Thermo-Electric Refrigeration

In 1834 a Frenchman by the name of Jean Charles Peltier discovered that passing an electric current through the junction of two dissimilar conductors caused it to either heat or cool the point of contact between them. A simplified peltier element is shown in Figure D49. The actual cooling process takes place within the element itself. There are no moving parts at this point, and no potentially damaging gases.

Considerable progress has been made in this area over the past twenty years, and thermo-electric cooling devices are readily available to both boatbuilders and the DIY public. The unit is simply let into a rectangular hole cut through the side of the icebox, and then connected to a 12V supply. A remote controller manages its activities in much the same way as previously described: When the engine is running it goes into 'rapid cool-down' mode (about 60W consumption), later reverting to economy mode (16W) with the engine idle. Again, there is a battery protection cut-off to avoid running down the batteries entirely.

12V thermo-electric refrigeration is really only suitable for smaller fridges – perhaps up to about 100 litres (3.5 cu.ft.), but 24V units can deal with up to double that capacity. Neither will make ice (except in dedicated gadgets designed for that purpose), but they will keep the contents of the cooler nicely chilled – quite enough for keeping food fresh and the beer in best fettle. We have one aboard *Spook* and it has done excellent service for many years.

Advantages claimed for thermo-electric cooling are:

- No moving parts, except for a pair of small fans to keep the air moving over both the warm and cool side of the unit. There is no actual movement of air between the two sides.

- Reliable in operation. As most of the device is solid state, there is none of the mechanical wear associated with conventional compressors.

- Compact size and light weight.

- The units can be quite easily retro-fitted into existing iceboxes.

- Very quiet in operation. There is virtually no vibration.

Feeding the Five Thousand

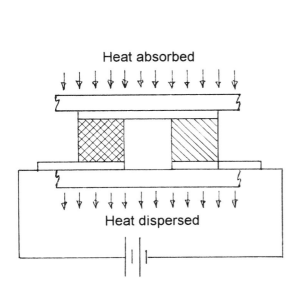

Fig D49 A Peltier element

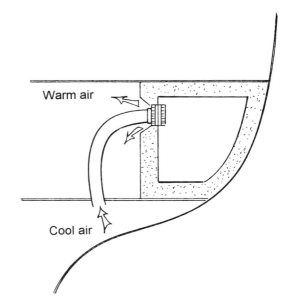

Fig D50 Drawing air from the bilge

- Environment friendly. There is no Freon to leak out or be disposed of when the unit is scrapped.
- Reduced electrical consumption.

However, the last claim requires qualification; it is probably only supportable in temperate or (at most) sub-tropical regions where there is not too great a difference between the ambient temperature and that required inside the fridge. Thermo-electric refrigeration units perform well at temperature differentials of up to about 25°C (77°F). Beyond that they begin to struggle and, in the most demanding tropical conditions, they will simply not be man enough for the job. Taking 5°C (41°F) as being an acceptable mean box temperature, you can appreciate that if the air inside the cabin rises above 30°C (86°F) – hardly an exceptional event in the tropics – your thermo-electric cooler will be hauling up the white flag. But, to some extent, this problem can be mitigated by drawing cooler air up from the bilge (Fig. D50)

So, it is very much a matter of choice, and that choice will be dependent upon such issues as: crew size, cruising area, electrical generating and storage capacity, type of cruising intended, and, of course, the initial cost of the gear. The needs of a singlehander undertaking a circumnavigation will be quite different to those of a charter yacht in the Caribbean or a man and his wife cruising the English Channel. Many people double-up on refrigeration equipment, using it as appropriate for different

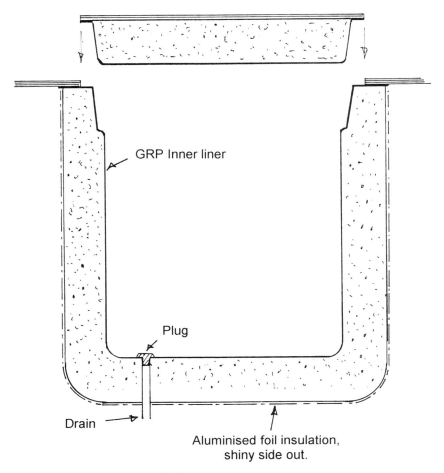

Fig D51 Fridge construction

circumstances. Luckily, the range of equipment is wide, and can be tailored to suit almost any application.

Insulation

No matter how brawny your refrigeration unit, its efforts will be wasted if it is being asked to chill a rickety container. If the box is badly sealed and inadequately insulated, the cold will rapidly seep away – though to think of it in those terms can be rather misleading. Cold is not an entity in itself, it is the absence of heat as dark is the absence of light. For the icebox, the enemy is really outside trying to get in, not the other way around. It is defence as much as containment that concerns us.

Figure D51 shows a typical installation. The interior liner is of moulded GRP with a thinner skin laid up over the insulation. For the boatbuilder, the most practical form of insulation is plastic foam. Of these, closed cell polyurethane is the best. This is available in sheet form of various thicknesses, or can be poured in situ by mixing two chemicals that foam and expand before setting. Sheet foam is to be preferred because of its consistent density and absence of major flaws, but this may be impracticable for irregular shapes.

Insulation thickness is the magic commodity – the more the better. A good rule of thumb is to use at least 1.0mm of thickness per litre of volume, with a minimum of 30mm (1.2in). In practical terms, this can be interpreted as:

Up to 50 litres (1.76 cu.ft.)	–	50mm (2in)
50–100 litres (1.76–3.5 cu.ft)	–	80mm (3in)
Over 100 litres (3.5 cu.ft. plus)	–	100mm (4in)

As a final refinement, on *Spook* I wrapped the outside of the insulated box with reflective aluminised film, shiny side outwards to repel heat absorption.

Ideally, a boat's fridge should be top-loading, with a hinged or lift-out lid that closes on seals (preferably double seals). For if cold itself is not an entity, then cold air certainly is. Being denser and heavier than warm air it will literally fall out of a front loading cabinet every time the door is opened. It will also flow out of any drain in the bottom of the box unless fitted with tap or water trap. Air is a highly penetrative fluid which will find its way out through the tiniest crack or fissure in your defence. There's no point in building a fortress if you then leave the door ajar.

Chapter 9
Plumbing

For long-distance cruising it was once thought generous to allot half-a-gallon of drinking water per person per day – with perhaps a little in reserve for emergencies. Many crews managed with less. In a nineteen-day Atlantic crossing two of us consumed a measly twelve gallons (54.6ltrs). But, in this as in so many other things, our expectations have increased, and at least double that allocation would now be considered the minimum.

Water is one of the most plentiful commodities on earth, and yet man has always been obliged to conserve it diligently – as if his life depended upon it; which of course it does. As Coleridge observed in his *Ryme of the Ancient Mariner*, the seafarer's position has always been especially ironic. Borne along on the stuff, often even drowning in it, and yet having to carry a comparative thimbleful to be eked out amongst the crew. In that regard, the sailor and the desert traveller are brothers.

The modern sailboat is often heavy on berths and low in tankage. Granted an eight-berth sailboat is unlikely to have a full complement aboard for a long passage, but the provision should be there to do so if desired. Invoking our gallon per person per day yardstick, a fully crewed eight-berth yacht with only 50 gallons (227ltrs) of water aboard could spend no more than six days at sea – not enough for a non-stopper from the UK to the Med, let alone a hop across to the West Indies. One designer told me that he was reluctant to encourage unnecessary weight aboard, but this is obviously ridiculous. Tanks are only heavy when they are full – and that is an option most skippers are qualified to exercise.

Ideally, the total capacity should be divided between at least two tanks. This guards against accidental contamination or water loss through leakage, and helps reduce the free surface effect when it all sloshes about. Personally, I prefer two separated supplies, each with its own filler, meeting at a Y-valve just upstream of the galley pump. This offers greater security than two tanks joined by a balance pipe where a failure in one could drain the other.

The exact position of the tanks is a matter for the designer to decide, but obviously they should be as low as is practicable, and as near to both the longitudinal and athwartships centre of the vessel as can be contrived. The 'standard' capacity can often be extended by fitting an extra flexible tank into a locker or other convenient space. However, I would be wary of using a flexible tank as the primary tank. They are vulnerable to abrasion and, as I mentioned in the introduction to this book, you could lose all your drinking water in a trice. But to supplement your basic tankage they can be very useful.

Water Filters

I have an early memory of cruising which touches not on good companionship or brave sunsets, but on the disgusting taste of the drinking water. It was almost taken for granted then that water aboard a yacht would be rank and fetid. For it is an efficient solvent, dissolving material from almost everything it passes over – the public pipeline, the supply to the pontoon, and the water tank and plastic pipes within the boat. From each it gains a little odour and flavour, and by the time the sailor gets to savour it, the resulting cocktail is invariably foul. Sailors once traded solutions on how to make it sweeter. Few were very successful, and we lived with its foulness in much the same way as we tolerated the leaky decks and primitive wet weather gear.

Now, thankfully, we need suffer no longer. A filter plumbed into the water pipe will remove all taints, leaving the water as pure tasting as if just drawn from a spring – better, indeed, than you might get at home.

The simplest filters are plastic or metal cartridges filled with activated carbon which is impregnated with silver to keep bacterial growth at bay. These are relatively inexpensive and many are disposable – though some have replacement elements which are fitted in permanent shells. Either way, the active bits should be replaced every twelve months, or sooner if a large volume of water has been treated.

Activated carbon filters do a remarkable job in removing tastes, smells and discolouration from water. They will also remove any chlorine which may have been added to the water (either at source or within the tank) to purify it, and most bacteria. They have very little effect upon the flow rate of the water, so there is no neck-bulging effort in a manually pumped system.

Offering a superior level of cleansing are more complex units which combine the activated carbon with microporous filtration, but these usually require more pressure (typically from 1.4 to 10 bar) than simple yacht systems can provide. However, they are commonly used on larger sailboats.

Desalinators

Osmosis is not a word to warm the heart of yachtsmen. Defined as the diffusion of a liquid through a semi-permeable membrane, it is the mechanism that imparts the

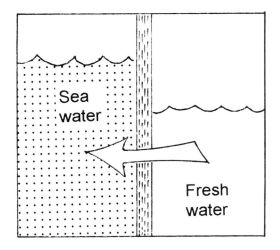 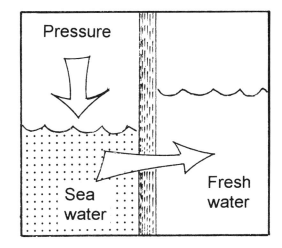

Fig D52 Osmosis   **Fig 53** Reverse osmosis

underwater acne to GRP hulls. In that context, the gelcoat is the membrane and the water outside is drawn through it to dilute the higher viscosity liquids – uncured resin, solvents, et cetera – contained in minute quantities within the laminate. And such is the power of this process, that enough pressure is built up under the gelcoat to form those fluid-filled blisters we all dread to see.

So, in response to thermodynamic law, when two liquids of differing viscosities are separated by a semi-permeable membrane (Fig. D52), there will be a transfer of liquid towards the most viscous side. But if sufficient pressure is applied to the more viscous side (Fig.D53) then the process will be reversed. With startling aptness, this is known as 'reverse osmosis', and its principles lie at the centre of the modern desalinator.

Figure D54 shows a simplified system. Water is drawn from the sea by a pump, which forces it under pressure (typically 800–900psi) through a membrane, folded so that a large area can be contained within a cylinder. Most of the sea-water passes straight through, and is discharged overboard again (though some systems ingeniously feed it back to the pump to provide a power assist effect), but a small proportion makes it through the membrane to emerge as fresh water.

There is a direct relationship between the power and capacity of the pump and the amount of fresh water produced. The larger units (in sailboat terms) use either 220VAC or 115VAC mains voltage motors, for which you would need a generator, or pumps belt-driven from the engine. Such desalinators might produce as much as 100 litres per hour (26gph). The limitations of a 12VDC supply obviously reduce the output, and here you could be looking at perhaps twelve litres per hour (3.2gph) – not exactly spectacular, but still a useful and sustaining source of drinking water. At the bottom of the pile are manually pumped units which might yield five litres

Plumbing

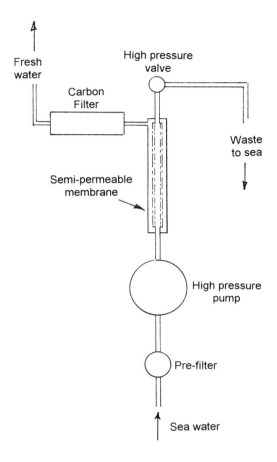

Fig D54 RO — desalinator schematic

per hour (1.3 gph), strenuously gained, and should really be considered as emergency equipment. Incidentally, most 12V pumps can also be operated by hand in the event of a power supply failure.

The RO desalinator is a relative newcomer on the sailing scene, and has yet to gain widespread acceptance amongst recreational yachtsmen who make mainly short passages. But for the long-distance sailor they have to be an unqualified boon. Not only are you spared the burden of carrying large quantities of drinking water – though it would be obvious folly to carry none – but you will know that whatever you drink is absolutely pure.

The Innovative Yacht

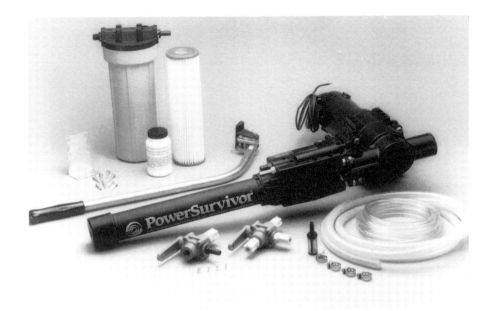

A Power Survivor 12V reverse osmosis desalinator kit.
The handle can be fitted for manual emergency use.

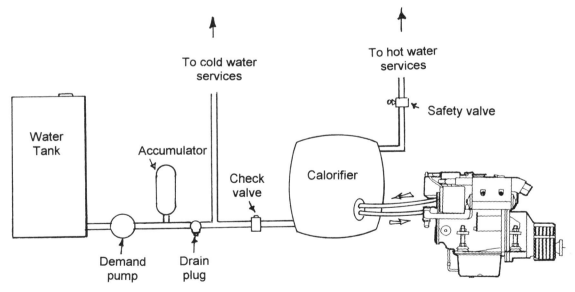

Fig D55 Cold and hot water supply

Plumbing

Water Heaters

I was strolling through the marina recently, when I came upon someone disembowelling his boat. Intrigued, I drew closer. As I watched, he staggered up the companionway, hugging an insulated hot water tank still festooned with pipes. He looked for all the world like a man fighting an octopus.

'Trouble?' I asked solicitously.

With a grunt, he tossed the octopus onto the dock and took to mopping his brow. 'Not any more', he told me grinning. 'Divorced the wife last week and the calorifier today. The two of 'em together could drink more than this old tub can carry.'

I nodded my understanding – at least about the calorifier. The temptation to make copious use of instantly available running hot water is a powerful one. Undisciplined crew members, apparently in the throes of some hysterical hygiene fixation, can empty the tanks as if punctured. For there is something about hot water which encourages the most extravagant liquid excesses. A pressurised cold water supply is bad enough . . . . but hot water! Sailors, more usually renowned for extreme parsimony with the provisions, suddenly start splashing it about like elephants at a mudhole. I suppose it must be the sheer joy of having the stuff on tap that turns both them and that steaming flow on.

And yet it would seem absurd to deny ourselves this luxury just for the lack of a bit of self-control. After all, on the average sailboat where the engine is run regularly,

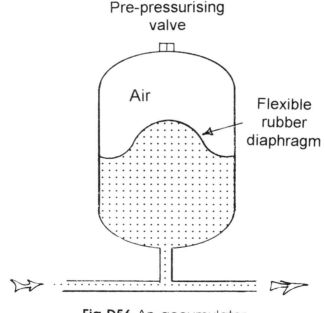

Fig D56 An accumulator

hot water is a by-product we can get for nothing, once the equipment has been installed. Most yachtsmen will appreciate the value of that.

A typical installation, for both a cold and hot water supply, is shown in Figure D55. When first switched on, the pump draws water from the tank and primes the system until the pressure reaches a pre-set level (typically 0.9 bar (13psi)), whereupon it switches off, ready for use. If any tap is now opened, the pressure will drop, the cut-in/cut-out switch will sense it, and the pump will leap into action to deliver the goods. The accumulator (Fig.D56) is an air filled container – sometimes divided by a diaphragm, but sometimes not – which allows some resilience to these fluctuations in water pressure, thereby smoothing out the jerky spurts of the pump.

Immediately downstream of the pump, the flow divides into the hot and cold water sides of the system. A one-way check valve should be fitted to prevent hot water flowing back into the cold.

The heart of the hot water supply is the calorifier. This is essentially an insulated heat exchanger. Within the outer jacket is a copper coil through which the hot engine coolant is pumped. The heat from the engine is thus transferred by conductance to the water held in the reservoir. Although calorifiers can be made to work (after a fashion) with raw water cooled engines, they are much more successful with indirectly cooled engines, where the coolant temperature is both higher and more constant.

An ingenious development takes advantage of the latent heat of melting. In addition to the heating coils, the calorifier contains an internal compartment filled with crystals which change to a liquid at 58°C (136°F), absorbing heat energy in the transition. As the tank cools, either by consumption or natural heat loss, this energy is released to re-heat the water – two tanks of hot water for the price of one!

It is also common to see 240V (115V in the USA) electric immersion heaters fitted to calorifiers. These are for use in port where a hook-up to mains electricity is available. Many calorifiers also have thermostatically controlled mixer valves, which add cold water to the output flow to give an even temperature and conserve the hot water.

Incidentally, strategically placed, a calorifier will convert any largish locker into a useful airing cupboard.

Boiler-type Water Heaters

More akin to the sort of hot water supply we are accustomed to at home are the active systems which function independently of engine heat or shore power. These are small boilers using either liquefied petroleum gas (butane or propane), paraffin (kerosene in the US) or diesel fuel drawn from the yacht's main tank.

Of these there are two distinct types:

1. Wall mounted, gas fired units, reminiscent of those in fifties bed-sits, which produce continuous hot water on demand. In these, the gas flame plays on a heat exchanger through which the water is passed. Water temperature is adjusted by controlling the flow rate through the heat exchanger. The slower the water flow, the longer it lingers in the heated coils, and the hotter the water. As the

water pressure is provided by the central pump, this type of boiler consumes no electricity. However, bottled gas is a relatively expensive form of fuel.

2. Remote boilers. These are often diesel fuelled units, tucked away out of sight in a locker or dedicated compartment. This type of boiler serves a closed circulating system which may also include radiators to warm the accommodation – very similar to a domestic installation ashore. Unlike the wall-mounted boiler, you will not get endless streams of hot water on demand, but the system is versatile, unobtrusive, and fuel efficient. On the debit side, electrical demand is high, as both combustion air and the water flow must be pumped by the unit itself.

Showers

When I was a youngster, freshly into sailing, and hanging on the words of the grizzled old tars I revered, I was told 'never deliberately let water into your boat unless you intend to drink it'. What these venerable codgers (who were probably at least in their thirties) would have made of the now commonplace yacht showers can only be imagined. For they are certainly the most profligate way of using water known. A five minute shower can consume thirty-five litres (9 gallons) – a week's drinking water for two on an ocean passage – grave-turning stuff for ancient mariners. But, in port or when coastal cruising from marina to marina, this is usually not a problem. And, the benefits (particularly to others) of being able to sluice down one's body, are indisputable. True, a bucket in the cockpit will do just as well, but there are climatic and other reasons why this is not always convenient.

Showers should be specifically designed as such. A curtained off corner of the heads compartment might grace the sales blurb and just qualify the arrangement as a shower in the eyes of the Trades Description Act, but is unlikely to work well in practice. Surprisingly, it seems not to have occurred to some designers that showers are places where water tends to get sprayed around a bit – actually, sprayed around a lot. To emerge pure and glistening from the deluge to find that you've effectively hosed down half the accommodation, is to know what it is to hate designers.

On the defensive front the two principal requirements are containment and collection. The first can best be achieved by building the shower into a dedicated cubicle – perhaps by sacrificing a superfluous heads compartment but, if not, then at least by locating the shower as far as possible out of the way of the toilet, lockers, and any other fittings which would not benefit from a drenching. Curtains should be both full length and full width, with the absolute minimum of gaps. Press studs can often be used to hold the edges closed, and it helps a lot if the drain pan has a higher than average lip. The run-off should go to a holding tank – never the bilge. Soapy water, combined with an agglomeration of discarded bodily bits, will turn any self-respecting bilge into a revolting slimy mess within a very short space of time.

Well worth the consideration is the 'sit-down' shower (Fig.D57), which can be tucked into a corner otherwise unusable for the purpose.

The Innovative Yacht

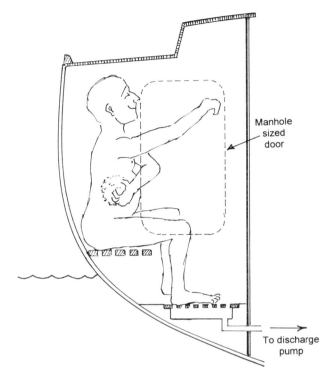

Fig D57 Sit-down shower

The Heads

In my experience, of all machines, marine toilets are about the most temperamental. This probably has something to do with their undeniably charmless lives, but they can be the very devil when they want to be. Perhaps they are simply getting us back for what we do to them.

It is an area where simplicity pays. My first sailboat had a bucket in a box and it never let me down once. Unfortunately, delicacy dictates that we do rather better than that. And, these days, so do environmental considerations.

Marine toilets pollute, and when a number of boats are clustered together – as they might be in a large marina – they pollute a lot, especially in those areas where there is no vigorous tidal exchange. Understandably, a weight of legislation is massing to prevent us being so anti-social.

Despite this, many boats are still built without holding tanks. However, under a European Community directive to be implemented soon all new pleasure craft will be obliged either to have a holding tank or some 'provision to allow the fitting of a temporary tank for use in those places where the discharge of human waste is restricted'. Such requirements are already in place in the United States, so we might as well get used to it.

A versatile arrangement covering all situations is shown in Figure D58. This allows for the toilet to be discharged either overboard or into the holding tank for proper disposal later. The workings of this system are principally controlled by the Y-valve at 'A', which will divert the waste wherever appropriate. An onboard pump allows the emptying of the tank, either overboard (where permitted) or into a quayside pump-out facility. And to keep your sea breezes fittingly bracing, the holding tank vent has an activated carbon filter to remove any odours and gaseous contaminates that might have escaped the influence of any added chemical deodorisers.

Treatment Discharge Systems

An ingenious alternative to the holding tank is represented by the Raritan Lectrasan. By applying an electric current to special patented electrodes, the salt (sodium chloride) in the water is used to create chlorine which treats and disinfects the waste.

The sequence works like this: The toilet's pump pushes waste into the first chamber of the unit, where it is reduced to small particles by rotating cutters. As the salt water in the system touches the activated electrodes, chlorine is produced. Subsequent flushes moves the effluent into the second chamber where it is mixed further and again is treated by the chlorine. After about four flushes, the treatment cycle has been accomplished and the now harmless residues are discharged into the sea. The chlorine, incidentally, reverts to its original state in the sea water.

Although this is an admirably elegant solution to waste disposal, there are a couple of snags: Firstly, each flush consumes about 1.5 amps of electricity – which could be significant on a fully crewed yacht – and, secondly, these units will not work in fresh (or brackish) water without a special salt feed.

Another system uses biological digestion, rather like a septic tank. Waste is flushed into the system where it is held by contained columns of redwood chips. There it breaks down primarily into water and carbon dioxide gas, before continuing into a chlorination chamber (to which the crew has added chlorine) which kills the bacteria. From there it is simply pumped overboard. If pumped manually, no electricity is needed.

All discharge treatment systems are inherently more complicated than holding tanks, and there is an associated susceptibility to mechanical failure. However, a strategically placed diverter valve can instantly convert the system back to direct overside discharge in the event of failure, or for use in open water.

The Innovative Yacht

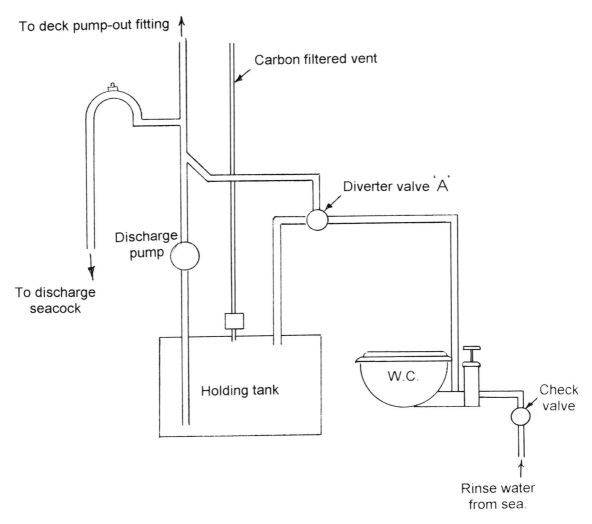

Fig D58 Heads, holding tank and overboard discharge

Chapter 10
Blow Hot or Cold

With the aid of the appropriate clothing (Chapter 15) and so long as we remain on deck, there is a great deal we can do to keep ourselves either warm or cool as the occasion demands. Anyway, it doesn't seem to matter quite so much when we are active and busy, the discomforts of the watch being part of what we expect of that fundamentally ill-advised occupation known as sailing. But as soon as we go below we become more vulnerable. The healthy rigours of the weather up top suddenly cloak us in misery as our adrenalin levels drop. And our memories start to play fondly with warm hearths and dry, inviting beds.

I recall with little pleasure a two-handed race around Britain. My boat had been hurriedly completed and there were numerous deck and window leaks. The weather was relentlessly awful – even a few splodges of sleet as we skated over the top of Scotland. Tricks at the helm were bad enough, but the snatches of rest we took below were grim beyond my worst imaginings. At the best of times the tiny cabin was as sparsely furnished as a penitent's cell. Now the interior was dripping, our clothing and sleeping bags soaked and, if it had not been for the pressing need to sleep away our exhaustion, absolutely nothing beckoned in the way of comfort and peace. It was not a pleasant way to go sailing. The truth is that all manner of hardship may be endured for short periods, but for the crew to 'make and mend' there must be a place of retreat; and if you cannot get ashore this has to be the cabin. To trade one awful environment for another will soon erode morale, whilst the reverse – the welcoming, restorative refuge – will lift the spirits and re-invigorate the body. When it is cold on deck it should be warm below. When it is blistering hot, it should be cool and airy. And it should always – but always – be dry!

Cabin Heaters

Given that a yacht is well designed and properly constructed, we can hope that it is possible to keep most of the sea and rain where it belongs – outside. What we are then left with is the cold, and to combat that we obviously need a heat source. Cabin heaters fall into two distinct groups: direct heating, where the unit relies simply on radiant and convected heat; and indirect heating, where the heat generator is installed remotely, and warmth is distributed throughout the yacht by air ducts or hot water pipes. Let's look at the options within each group:

Direct Heating

These are the simplest, and often the most reliable ways of heating a space, but their usefulness is limited to the smaller sailboat where natural dispersal of the warmed air is considered sufficient. Clearly, to create heat we have to burn something. And, in the case of direct heating systems, the choice of fuel is the widest. None make any demands upon our electrical capacity.

Solid Fuel

For the traditionally minded amongst us, there are still stoves of the pot-bellied persuasion to be had. But, whereas these might look dandy in a Bristol Channel pilot cutter, or a Grand Banks schooner, they are unlikely to add grace to the interior of the modern sailboat. For those who think that large lumps of cast iron should only be used as ballast, there are more refined variants on the theme – usually of polished brass or stainless-steel – which do the same job in a rather less monumental manner. The largest solid fuel stoves will burn almost anything, but for the smaller modern units charcoal is the best choice. It burns cleanly with little smoke or ash and gives off a comforting and agreeably dry heat. Thanks to the universal popularity of barbecues, it is available virtually everywhere. But, although light in weight, it is too bulky to be truly convenient, and can create an unbelievable mess if it gets out of control. Always store it in plastic tubs. Paper sacks or cardboard boxes may turn to pulp and your locker will look like a Siberian coal-face before long.

Paraffin (Kerosene) Oil

These systems use a manually pressurised tank which feeds fuel to an 'Optimus' type burner similar to those found on cookers. And, in the same manner as cookers, the burners require pre-heating with a couple of spoonfuls of alcohol (methylated spirits). The combustion of any hydrocarbon will produce large amounts of water vapour which, if it is not to condense out into the cabin, must all be vented to the open air through a small diameter (typically 28mm(1.1in)) flue. Obviously some heat is lost in the process, but at least that which remains will be dry. And, paraffin is an inherently smelly fuel which some people find objectionable. Very minor leaks, and partially blocked burner jets (which will prevent complete combustion), can soon have the cabin smelling like a refinery. Proper maintenance is the answer here. Keep the gear in good working order and the only evidence of its existence will be glorious heat! Where this type of heater is concerned, I am something of an enthusiast. I have been shipmates with them for years and have had many reasons to be grateful for the comfort they provide. They are relatively inexpensive to buy and astonishingly miserly with their fuel. For the smaller yacht, they have my vote.

Diesel Oil

These look outwardly similar to paraffin heaters but there are essential differences within. Diesel heaters use 'drip-pot' burners which are fed from an unpressurised gravity-feed tank. The flue must be of larger diameter than a paraffin heater – about 60mm (2.3in) being typical – which rather detracts from the neatness of the installation, particularly on smaller sailboats. These units will also run on paraffin, but there is an obvious convenience in having a heater which drinks from the same well as your engine. Perhaps this is their only advantage. In every other regard, the paraffin unit is usually the better choice.

Liquefied Petroleum Gas (LPG)

In my youth I once signed on for a late autumn trip aboard an elderly carvel gaffer owned by a sailor I greatly venerated. The equinox was some weeks behind us and the days were getting short. After an uneventful crossing from Guernsey, we had dropped our overnight hook in Newtown Creek prior to making the final few miles to her home port, Hamble. You could tell by the still, clear feel of the evening that there was going to be a frost by morning. 'By gum', my skipper said, coming below, 'I believe it's chilly enough for Old Smokey'. He plunged into a locker and soon emerged clutching a rusty cast iron ring, peppered with holes and trailing a mouldering length of garden hose. From another locker he produced four house bricks which he handed to me one by one. I watched hypnotised as he placed the ring on an upturned bucket, stacked the bricks on top of it, forced the hose over a gas outlet, and then struck a match. With much popping and hissing, the thing was alight, a fiery volcano that licked upwards between us in the low cabin. It all looked incredibly dangerous but my companion was ecstatic. He clapped his hands gleefully. 'Soon be as warm as toast,' he chortled. 'Got it from a boatbuilder pal of mine. He used it for melting tar!'

Thankfully, today's LPG heaters are considerably less fearsome. They fall into two types. Some units still use naked flames, and look very similar to the paraffin heaters previously discussed, and are vented in the same way (via a flue) to the open air. But gaining in popularity are the catalytic type. These are fairly cheap to buy and are neat and compact. They burn without any flame at all and at a relatively low temperature, thereby minimising the risk of an accidental fire. LPG is not a poisonous gas and its products of combustion are harmless, given adequate ventilation.

When gas heaters (or, indeed, any other appliances) are working normally, they produce quantities of carbon dioxide (CO_2) and the usual water vapour – neither of which are cause for concern. Provided that ventilation is adequate, the balance between the oxygen in the air and the combustion by-products remains stable and safe. However, if the ventilation is restricted – perhaps a partially blocked vent or flue – then three things could happen: 1. The oxygen level in the cabin decreases. 2. The carbon dioxide level increases. 3. As the result of 1 and 2 the appliance becomes potentially lethal as the flame characteristic changes and it starts to produce harmful quantities of CO_2's deadly cousin, carbon monoxide (CO).

Now, as well as being poisonous, carbon monoxide is also colourless and odourless. Concentrations can reach (and have at times unfortunately done so with tragic results) dangerous levels without detection. All unattended LPG appliances (eg. cabin and water heaters) should be fitted with devices which not only protect against flame failure but also against an increase beyond a safe level of CO_2. These are known as Atmosphere Sensing Devices or ASD's, which usually take the form of a specialised pilot burner which controls the appliance's gas supply via its own flame failure valve. When the atmosphere becomes polluted with CO_2 the pilot flame 'lifts' off the burner, thus cooling the thermocouple which, in turn, shuts down the gas supply.

When you are out shopping for gas appliances, you would be wise not to assume that everything you see will be properly protected. To meet the relentless public demand for low-cost goods, the specifications of some are cut to the bone. But CO poisoning is not to be relished – especially on a sailboat where you might be far from medical assistance. On entering the lungs CO impairs the blood's ability to absorb and carry oxygen to the point where death can occur. And, of those who survive a severe dose of CO poisoning, many are left disabled with irreversible brain damage.

Indirect Heating

Obviously the warming effect of direct heaters is extremely localised, and really only suitable for the sort of compact lay-out you might find on smaller sailboats. If the accommodation is divided and you want an even spread of heat, then we must climb the sophistication ladder to central heating installations.

By far the most powerful systems use ducted hot air, blown from a fan heater fuelled by gas, paraffin, or diesel. The actual heating of the air is indirect. A fan blows fresh air over a heat exchanger, inside which the fuel is combusted. The combustion gases are then vented through an exhaust fitting to atmosphere, whilst the heated fresh air is blown through flexible ducting to strategically placed outlets. The operation of the unit is controlled by a thermostat and the outlets can be opened or closed to suit the needs of each compartment. A 2KW, single outlet unit would comfortably serve a twenty-five footer (7.6m), and it would need about 12–15KW, and perhaps as many as eight to ten outlets, to do similar service to a yacht of, say, twenty-five metres (82ft). Electrical drain is obviously related to output, but, in relative terms, can be high for a sailing yacht with its limited generating capacity. Typically, a small unit would draw about two amps on its low setting, whilst a mid-size one, suitable for an 11m (36ft) yacht would be pulling about 3.5 amps from the batteries. These are fairly expensive pieces of kit, and they have unfortunately gained a reputation for being mechanically temperamental. But, in those cold winter evenings, if you want to boost the cabin temperature to blissful levels quickly, they are pretty hard to beat.

As we mentioned earlier in this chapter, water-filled radiators are also a possibility, being served by the same remote boiler which supplies the domestic hot water. The radiators themselves can either be passive, relying solely on convection, or can be fitted with fans – rather like car heaters – to improve circulation of the warmed air. This last may be useful in hard-to-heat areas such as wheelhouses, but will obviously add to the already significant electrical consumption.

The principal arguments in favour of this form of heating are that hot running water and cabin heating can be gained from the same source, and that the associated pipework is neater, less intrusive, and less susceptible to damage than air ducts. The counter view centres on the relatively high equipment costs and the extra weight of a water-filled system. On balance, this type of heating is impractical for the smaller yacht but gains merit with size.

Air Conditioning

Air conditioners are basically refrigerators through which air is first blown, then to be directed onwards into the space you wish to cool. But, as we covered in Chapter 8, an essential requirement of an efficient refrigerator is good insulation. Boats are spectacularly deficient in this quality, which gives some indication of the thankless task facing any marine air conditioner. They consume electricity – lots of it. To keep the interior of the smaller sailboat artificially cool is difficult. Larger yachts, with mains voltage generators, can handle the electrical demands, but this is rather outside the scope of less grand craft. Granted there are 12VDC units on the market, but their appetites are still voracious, quite beyond the charging capability of a boat under sail. But there is a role for small boat air conditioning and that is alongside. Marinas can be stifling places and boats are tiny enclosures in which heat levels can soar unbearably. Supported by shore-power, the impracticable becomes possible.

Portable air conditioners are probably the most popular amongst yachtsmen. They can either be free-standing (though these are invariably shore voltage units) or are mounted in any convenient hatch – usually the forehatch or cabin skylight.

Air conditioners are especially useful in humid climates. For, not only do they chill the air, but they also remove moisture from it. Where humidity is a problem, it is not unusual for owners to keep their boats permanently air conditioned alongside, to hold mildew and other fungal horrors at bay. But, for the average offshore sailor, the delights of refrigerated air remain an impracticable luxury.

Chapter 11
Engines

Times have changed. It used to be that the yacht auxiliary was regarded as an object of contempt – almost embarrassment – a shameful thing to be hidden away like some idiot son consigned to a discreet but distant institution. And to be seen using it was somewhat akin to being caught cheating at cards. 'Saw you coming down the Wallop last evening. Had the iron tops'l set, did we, old chap? Well, I suppose you wanted to get home for your tea.' It was a device everyone had but nobody mentioned, unless disparagingly – and certainly nobody loved.

No longer. By comparison, today's yachtsman is a shameless soul. Not for him the hours spent milking precious fractions of a knot out of listless air, or the nail-biting heavy weather charge to his mooring under sail. Now, if the wind falls still or there is some awkward manoeuvring to do, it's on with the engine without a second thought. And quite right too. To enjoy for a while the somnolent pleasures of being becalmed is one thing. To endure it unnecessarily when you have an alternative means of getting you home is absurd. I recall a miserable twelve-day crossing of the Gulf of Mexico, a dead and useless engine down below and not enough breeze to twitch the whiskers on a cat. Given a half decent auxiliary and sufficient fuel, it would have been – and should have been – a four-day trip at most.

And it also used to be that petrol engines – many of them primitive two-strokes – were the preferred choice on small to mid-sized sailboats. But over the past several years, immense efforts have been invested in the development of small diesel engines and now, thankfully, the marine petrol engine, with its vulnerable, unreliable electrics and dangerous fuel, is virtually obsolete. Recognising this, from here on in this chapter when we talk of engines we mean diesel engines.

Engine Size

When Eric and Susan Hiscock commissioned *Wanderer III*, they opted for a 5hp Stuart-Turner two-stroke. They rationalised that, when ocean voyaging over vast distances, the principal purpose of the engine would simply be to get the boat into and out of port, and you don't usually need a lot of horses for that. Later they acknowledged that they had rather skimped in this department which, by modern standards, they clearly had. A thirty-footer of fair displacement propelled by only 5hp would obviously perform poorly in anything other than the lightest conditions – and would probably be brought up absolutely short by anything serious in the way of wind and waves on the nose.

But, in those days, they had an excuse. Then, as now, weight and horsepower were directly related, but rather more extremely so then, than today. The inducement for owners to save weight – either for sailing performance or payload reasons – by installing the smallest possible auxiliary was a compelling one, frequently resulting in grossly underpowered installations. For example, to obtain a measly 10hp from a diesel you might have been looking to fit a lump weighing perhaps 200kg (440lbs). By contrast, for that weight penalty now, you would expect to get at least 40hp from a modern engine, and could get your 10hp (if that really is what you want) from an engine weighing only about 80kg (176lb).

Also, marine diesels were, in relative terms, very much more expensive than they are now. Naturally, with the acceptance that engines really are essential items, and that every sailboat should be able to make robust and significant mileage under power in nearly all conditions, the market has risen to meet these needs. Consequently, there is absolutely no reason for any boatbuilder to underpower his products – and happily, few do.

There are various formulae for determining the ideal engine size, but this decision is usually made quite satisfactorily on an empirical basis. The important thing is to have enough power on hand – definitely better too much than too little.

Living With the Beast

Yacht owners rarely have much choice in the matter. Most boats come with the engine already in place, and you are stuck with it. And, although some would disagree, so long as the engine is of the right size they will probably be satisfied. The difference in quality between one brand of engine and another is usually insignificant – and all are very reliable. For the diesel auxiliary is often a marinised version of a transport or industrial engine which, in landlubberly role, is built to deliver thousands of hours of service before replacement.

The yacht engine by comparison leads an idle life. A hundred hours a year would be considered strenuous – and most do a lot less. At that rate it would plainly take a very long time to wear the thing out – perhaps even longer than the boat itself. Marine diesels usually die of neglect, not overwork.

But if the engine is well made, what about the installation? Here, I am sorry to say, the story is not so encouraging. There is not much a yacht builder can do to lessen the unit price of the engine itself, but he can certainly cut corners when he fits it to the hull. When considering the total auxiliary power package – the engine, its bearers and mounts, the exhaust, the prop and shaft assembly, and the fuel supply system – you will find that most standard installations are about as basic as you can get. If you want it done properly, you may have to do it yourself.

Noise

Ask sailboat owners what they think about motoring, or even motor-sailing, and most will shudder in response. Of course, by definition they would prefer to be sailing – otherwise they would have chosen a power-boat in the first place – but their emphatically negative reaction smacks of more than just plain old disappointment. Press them a little further and they will tell you that what they detest about it is the noise.

And they are members of a large fellowship. The view that a sailboat under power is inevitably noisy is widespread. And, as we know, many are – some thunderously so. On one sailboat I can think of, it is impossible to converse in the cockpit without shouting, and the cabin is quite unbearable – a cacophonous hell-hole in which you can stay for no more than a few minutes without being in serious need of aspirin.

But, how absurd! A sailboat engine is a tiny thing compared to that fitted to a bus, or even a large car. The notion that it should be able to invade our peace so intrusively would be unthinkable in any other context. It is like being adrift with some untamed brattish child that shouts and screams its will across our lives. Personally, I find both intolerable.

Consider it in principle. A violin is basically an empty box which resonates to a source of vibration, namely its strings. On its own, the strings would have a puny voice, barely audible, but allowed to amplify in the voids and sounding surfaces of the fiddle and it becomes authoritative enough to fill a concert hall. Well, a hull is also a box which is largely empty and it too has a source of vibration – a very strident source. Now you would expect a fiddle maker to do his utmost to ensure that every little quiver, with all its colouring harmonics, would be transmitted throughout the body of his instrument. The size and the quality of the tone will depend on it. But a boatbuilder? Surely his interests should lie in the opposite direction.

Quite naturally, any machine whose very function depends upon a succession of explosions is going to be essentially noisy. But there is no reason to encourage it. Sound is carried through the air in the form of waves (actually variations in compression of the air), and along solid objects by vibration. In order to tame the beast, we must strive to do two things:

1. Limit the spread of vibration by introducing flexible links between rigid elements. This includes the engine itself, the controls, the propeller shaft and the exhaust.

2. Physically block the passage of airborne noise – a fragile creature, readily discouraged. For sound to travel through, let's say, a wall the air compressions on the source side must be converted to vibrations within the wall, and then back into air compressions on the other side. Obviously a great deal of energy

Engines

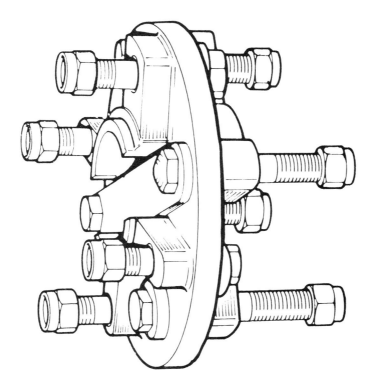

Fig D59 R & D coupling

is absorbed in the process, and the thicker the wall and the less readily it vibrates, the more energy loss there will be.

So, if this is what is desirable, how does the average sailboat score? The engine is bolted to softish mounts but that is about the only real concession to noise reduction. Luckily, there are devices and materials on hand to improve the situation.

Flexible Shaft Couplings

These fit between the prop shaft and gearbox half-couplings, and help reduce the transmission of vibration down the shaft. There are various designs available, but the one shown in Figure D59 is fairly typical. In this example, a specially contoured moulded polyurethane disc is able to flex freely, thus taking up any minor misalignments which arise as the engine vibrates on its mounts.

The fitting of a flexible coupling is an inexpensive, but useful improvement on the 'basic' installation.

The Innovative Yacht

The Aquadrive constant velocity shaft coupling in *Spook*.
Note the strong transverse member which bears the propeller thrust.

Constant Velocity Joints (CVJs)

These do the same job and more. Whereas the simple flexible coupling will only tolerate minor misalignments between engine and shaft, the CVJ can deal with gross, often deliberate, differences between them – typically to as much as 16°. A hefty thrust-bearing is also incorporated which, with the CVJ mounted on its own cross-bearer, takes the forward (or aftward when going astern) push from the propeller. Thus relieved of the task of resisting the propulsive thrust – indeed, now almost entirely isolated from the stern gear assembly – the engine can be carried on much softer mounts, with some additional reductions in transmitted vibration.

Engines fitted with CVJs and stood on softer mountings can actually reduce vibration by up to ninety-five per cent and structure-borne noise by fifty per cent – no mean achievement. And, although they are quite expensive bits of kit, their cost will be mitigated over the years by less wear to the gearbox, shaft, stern gland and bearings.

Exhaust Bellows

Rubber exhaust hose also transmits vibrations which will find their way to the hull unless suppressed. On wet exhaust systems, synthetic rubber bellows can usefully be inserted at various points to reduce this effect, and to minimise the risk of damage to rigid components such as silencers, water traps, or elbows. How many and where they are most profitably fitted is usually dependent upon the configuration of any particular installation, but the most useful site is usually on the engine side of the silencer.

So much for vibration. What remains is the engine's voice, emanating from the block itself and also being conducted down the exhaust, usually to within a few feet of where you sit in the cockpit.

The Silencer (or Muffler)

Most sailboats have a 'wet' exhaust system where, after having done its bit circulating through the heat exchanger (or the engine's various galleries in a raw water cooled engine), the cooling water is injected into the exhaust just after it leaves the manifold. This rapidly cools the exhaust gases, which reduces their volume and stifles much of the sound. It also allows the use of such materials as rubber and plastic, which would burn or melt if the gases were at their full heat.

When the engine stops, the water within the length of the exhaust runs back to collect at the lowest point. It is clearly important that the exhaust system be arranged so that water cannot re-enter the engine, and the need to safeguard against this also gives us the opportunity to further reduce exhaust sound.

Figure D60 shows a water-lift type silencer, and Fig D61 the same silencer fitted to a typical sailboat installation. The silencer can be made of plastic, stainless-steel, or filament wound GRP. Perhaps surprisingly, stainless-steel is the least reliable, being prone to leakage at the welds. Although the most expensive, GRP is probably the best choice.

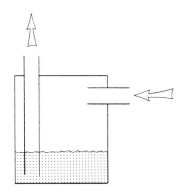

Fig D60 Water lift silencer

The Innovative Yacht

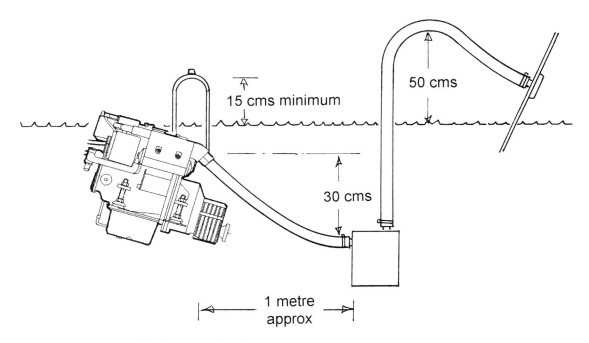

Fig D61 The water lift silencer fitted to a boat

Although not absolutely critical, the best location for the water-lift silencer is about one metre (39in) abaft the engine and at least 300mm (11.8in) below the injection point. To keep it reasonably unaffected by heeling – which could alter its height relative to the engine – it also should be on or near the centreline.

It can be assumed that about ten per cent of the total exhaust hose volume is water, and as a rule of thumb the silencer should have three times the capacity of that water content. To put it another way, with the engine stopped, the run-back of water should fill the silencer by no more than a third.

Acoustic Insulation

And, lastly, the most powerful weapon against airborne sound. All sorts of absorbent materials are sometimes used – rock wool, polystyrene tiles, polyethylene foam, I have even seen egg cartons – but by far the most efficient are those barriers which have been specifically developed for the purpose.

Figure D62 shows the construction of a dedicated noise insulator. These can be obtained in a variety of thicknesses, and I am sure it will come as no surprise to learn that the thicker the better.

As far as possible, the engine should be totally encased with insulated panels and there should be the minimum of bolt-on accessories (fuel filters, electrical equipment,

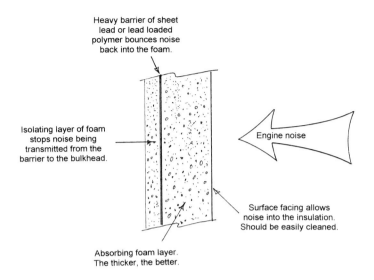

Fig D62 Noise insulator

et cetera) fitted to those bulkheads to provide uninsulated paths to the world outside. Seams between adjoining sheets of insulation should be carefully taped over, not only to close any gaps but also to guard against a dislodged piece falling in towards moving parts.

The best quality insulators – and these should be the only ones you choose – are fire resistant and will provide a barrier against an engine fire, retarding its spread and allowing the (hopefully automatic) fire-fighting equipment to do its stuff.

So far as noise suppression is concerned, nearly all engine installations can be greatly improved, and some completely transformed. Many people become accustomed to the roar from below, and simply minimise their pain by using the engine as seldom as possible. This is a pity when some fairly simple adaptions will make the lump a friendly beast to live with.

And Stoking It

In most regards diesel engines are extraordinarily forgiving things. The one in our yard launch spends its life outdoors, covered by nothing more than a leaky wooden box – employed, it has to be said, more to screen it from public gaze than to give it worthwhile protection. Visually it is not impressive. The paint has long since gone, and has been replaced by a leprous coating of rust that sheds its scale into the bilge

The Innovative Yacht

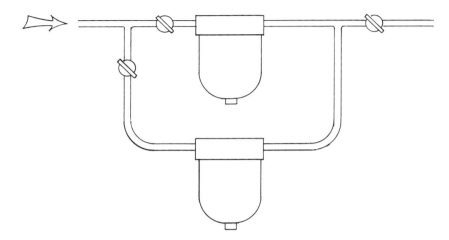

Fig D63 Fuel filters in parallel

like brown dandruff. And yet it runs, and runs well – often to the amazement of strangers who marvel at it as they might a geriatric athlete.

But in one vital respect it is temperamental. Shabby and decrepit though the machine might appear, it has refined tastes and will not tolerate anything dubious in the way of fuel. And neither will most of its breed.

Once, in a foreign port, I tanked up with fuel, drawn from a large steel tank on the quay. Later, we motored out into a flat calm and were well caught by the tide when the engine stopped. Investigation revealed the pre-filter to be blocked, and the fuel contaminated with some unidentified sediment that hung in suspension. In one sense we were lucky. We had cleared the dangerous navigation hazards and were out into deep water. Twenty minutes earlier and it would have been a different matter. Then we would have had insufficient time to change the filter element before being swept down onto the rocks.

Now we have a second pre-filter connected in parallel with the original one (Fig D63). Should the first become clogged, we can bring the other one on line in an instant. Of course, that too could soon choke, but at least it buys us precious time to sort ourselves out.

A better arrangement would be to have a small day tank – which could contain no more than perhaps five litres (1.3 US gallons) – filled from the main tank via an intermediate filter. On many boats this could be positioned so that the fuel transfer was done by gravity, but, if the main tank is fitted low in the hull, a hand or electric pump might be necessary. Either way, many standard fuel supply systems can quite easily be modified so that you can breathe a lot easier after buying fuel from doubtful sources – a matter in which the cruising sailor sometimes has little choice.

Chapter 12
Electrics

Once, crossing the Atlantic in 1974, the only electrical equipment I had aboard was an echo-sounder, a tape cassette player, a radio receiver, and a couple of flashlights – all powered by internal dry-cell batteries which I replaced when they became exhausted. Navigation and cabin lights were wick-type paraffin lamps, backed up with a pressurised Tilley lantern as a last resort 'ship scarer'. When I think of it now, I wonder at the primitiveness of it all, but by the standards of the day there was nothing very unusual about it. Small sailboats, embarking on long passages, simply did not have the means of supporting sustained electrical consumption. Many ocean voyagers therefore chose to do without it altogether.

By contrast, my present boat bristles with electrical gear and, again, there is nothing very odd about that. For we have become more and more dependant upon electricity, and technical developments in this field have made it so much more practicable. The sea is an exacting environment for any kind of circuitry, but modern materials, encapsulated micro-electronic components, and generally dryer hulls have gone a long way to making the equipment more durable and dependable.

Electrical systems must be viewed in their entirety. There is little point in attending to one aspect of it whilst ignoring another. Consumption must be matched against generating and storage capability. At the end of the day it all boils down to a simple energy equation: You can take out only as much as you put in. Not for nothing are multiple batteries called 'banks'.

The interaction between supply and demand can be evaluated by considering the following:

- The typical power consumption of the entire system. This can either be expressed in watts or amperes (A). Amps is probably easier as it can be related directly to amp-hours, the most usual method of defining a battery's storage capacity.

- Storage capacity in the form of batteries.

- Generating or charging capabilities from whatever source – alternators, wind generators, solar panels, etc.

- Voltage regulation to provide efficient control of the charging systems.

In this chapter we will tackle these factors in approximately that order but, before we become too bogged down in the technical mire, some clarification of the electrical units used might not go amiss. Those with a technically superior education can, of course, skip this bit.

Volts (V): Named after Allesandro Volta who constructed the first battery – the 'voltaic pile'. A volt is a unit of electromotive force and can be loosely thought of as electrical speed or pressure.

Amperes (A): This unit honours the French physicist, A. M. Ampere, who died in 1836. Nearly always shortened to 'amp' verbally, this is a unit of current, and can be thought of as a quantity of electricity.

Watts (W): The Scottish scientist James Watt lends his name to this unit of power which is the combined result of volts and amps and represents the power of the current – in other words, the working capability available.

The oft-flayed analogy compares the behaviour of electricity to water flowing through a pipe. The speed of the water through the pipe represents volts, the quantity of water is amperes, and the useful energy of the water spurting from the nozzle is obviously a combination of both speed and quantity and can be likened to watts. Using this model, it can easily be understood that a small quantity (amps) of water would need to move faster (volts) to do the same work (watts) as a larger quantity moving slower.

More formally, the relationship between the two units can be expressed as:

$$W = V \times A$$

And, transposing:

$$A = \frac{W}{V}$$

How Much Electricity Do You Need?

This is the crunch calculation which every sailor should make, but many avoid. Before you can evaluate your charging requirements, you must first determine a predicted figure for consumption. Many owners run their electrical systems at a deficit – relying on frequent transfusions of shore power to make up the shortfall – but the long-distance cruiser, with his need to be self-reliant, must be rather more circumspect.

In order to do this you should first define a representative style of cruising for yourself, and then tot up the amount of electricity each item of equipment is likely to consume over any twenty-four hour period. You will need to know something about the electrical greed of each item. Those cabin (and other) lights will probably be rated in watts, so you will have to convert them to amps by dividing the wattage by the boat's system voltage. For example:

For a 25W masthead tricolour light : 25 ÷ 12 (volts) = 2.08A
Say, 2A in round figures.

Below are listed typical power consumption figures for the kind of gear you might find on a mid-sized sailboat. For some of the more powerful items – autopilots, refrigerators, windlasses – there can be a very wide spread between types and brands, so you should check the specifications for the actual units on your own boat.

Electrical Loads at 12V	Amps
Anchor light	1
Autopilot	1–10
Bilge pump (submersible)	2–12
Cabin light (tungsten)	0.5–1
Cabin light (fluorescent)	1
Cabin light (halogen)	0.5
Depth-sounder	0.1
GPS	0.1–1
Fresh water pump	6
Log (knotmeter)	0.1
Loran	1–1.5
Masthead tricolour light	2
Radar	4–8
Refrigerator (compressor type)	5–8
Refrigerator (thermo-electric)	5
Running lights (side & stern lights)	3
SSB radio (receive)	1–2
(transmit)	5–30
Steaming light	1
VHF (receive)	1
(transmit–full power)	6
Windlass	15–150

Of course, most equipment runs only discontinuously, so we have to estimate an operational time for each item in order to calculate how much power it will draw during our specimen day. Some equipment – GPS, or radar, for instance – may be left on permanently whilst on passage, and then closed down entirely when in port or at anchor. The windlass will see no use at all when at sea, but may be working its ravenous heart out in an island hopping, lunch hooking jaunt through the Balearics or Bahamas. In higher latitudes in mid-summer, the short nights will place only scant demands on the navigation lights. Six months later in the same waters, the seasonal gloom could pose a major strain on your electrical resources. There are no standard circumstances. It will be up to the individual skipper to set his own terms of reference and, wisely, to play safe and err on the pessimistic side. No matter how miserly you might think you will be with the electrics, in reality we all tend to be more careless than we ever intended.

As an example, let's take a fairly standard family cruiser enjoying the delights of the Atlantic coast of France. It is around the date of the summer solstice and the nights are at their shortest. Traffic is heavy and the navigational duties around that rocky coastline are, as always, demanding. The itinerary includes short day hops and a longer passage at sea.

At Sea:

Equipment	Rating	Hours Used	Total Load
Tricolour light	2 amp	6 hours	12 amp-hours
Cabin lights (2)	1 amp each	2 hours each	4 amp-hours
Auto-pilot	2 amp	10 hours active	20 amp-hours
VHF radio	0.8 amp average	24 hours	19 amp-hours
Instruments	0.4 amp	24 hours	10 amp-hours
Radar	5 amp	3 hours	15 amp-hour
Refrigerator	5 amp	8 hours active	40 amp-hours

TOTAL: 120 amp-hours

The Same Boat at Anchor:

Equipment	Rating	Hours Used	Total Load
Anchor Light	1 amp	6 hours	6 amp-hours
Cabin lights (4)	1 amp each	3 hours each	12 amp-hours
Refrigerator	5 amp	10 hours active	50 amp-hours
Radio, stereo etc.,	1 amp	4 hours	4 amp-hours

TOTAL: 72 amp-hours

From the above it can be seen that this sailboat is more electrically extravagant on passage than when at anchor. This is hardly surprising with such power-hungry lumps as the autopilot and radar in use. Of course, electricity can be conserved by using these devices sparingly, but the fact remains that the modern yacht is designed to be reliant upon its electrical and electronic systems, and that demand will be likely to remain high despite reasonable economies. We need to know how to accommodate this demand.

Batteries

Boxes of electricity, my son, Angus, would call them – and not a bad description at that. And the bigger the box, clearly the more you can carry in it.

Battery capacity is expressed in ampere-hours (Ah). A 200Ah battery would therefore obviously have double the capacity of one of 100Ah. This ampere-hour rating is derived differently in the UK and the US, which raises anomalies when making direct comparisons. To understand this difference is to appreciate that, when talking of battery performance, all is not quite as straightforward as it might seem. Let's assess the two rating methods first:

- Batteries in Britain are rated over a 10 hour period. A battery capable of sustaining a discharge of 10 amps for 10 hours is deemed to have a capacity of 100Ah.
- In the United States, the rating period is over 20 hours. There, a battery discharging fully at 5 amps over 20 hours is rated at 100 amps. However, should the discharge rate be increased to 8.4 amps, that same battery will be dead flat in only 10 hours – making it an 84Ah battery by UK standards!

Of course, the detrimental effects of high discharge rates on battery capacity is not just a chauvinistic penalty visited on Americans. All batteries, whatever their origin or rated capacity, will perform better if discharge rates are low, and worse – possibly much worse – if discharge is high. For example, if a 100Ah (UK rated) battery was subjected to a discharge of 30 amps, it would be flat in about two hours – effectively reduced to 60Ah, or sixty per cent of its rated capacity. Nominal ratings are, therefore, not to be taken too literally. They should be considered within the context of the system as a whole.

Battery Types

Lead-acid yacht batteries should always be of the deep cycle type – sometimes misleadingly known as 'traction' batteries because they often do service in golf carts and powered wheelchairs. Automobile batteries are virtually useless for the purpose. These are designed to deliver short bursts of electricity to crank and start the engine; and thereafter languish with their feet up whilst the alternator does all the work. Sailboat batteries lead miserable lives. They are subjected to haphazard recharging regimes, and are then discharged to levels which would rapidly kill a car battery. The successful marine battery is rugged out of necessity.

The feature that distinguishes a deep cycle battery from its wimpish counterpart lies in its internal construction. The chemical action in the positive plates involves converting lead dioxide to lead sulphate and back again. This conversion shakes loose the bond between the paste-like material and its supporting grid, and the more profoundly the battery is discharged, the greater this effect. With every discharge cycle, material falls away, thus reducing the overall capacity of the battery and 'silting' up the bottom of the battery container with a metallic sludge. Should the level build up to touch the bottom of the plates, they will be shorted out and the battery will fail entirely.

Deep cycle batteries use stouter, thicker plates, and denser active materials. They also use elaborately constructed separators to keep the plates apart and the active paste where it belongs. Even the best of these will shed some material over the years but, if treated half decently, this loss will be far less than would be expected from a cheap car battery. This is very much an area where you get what you pay for.

Alkaline Batteries

Also known as nickel-cadmium or ni-cad batteries. These almost indestructible batteries can have all manner of abuse heaped upon them without complaint. They can be deep cycled, left discharged, and even dead short circuited without harm. Twenty to thirty years of useful life is not unusual for this type.

But, unfortunately, we still live in an imperfect world, and ni-cads have their down side. Amp-hour for amp-hour they are bulkier and heavier than the lead-acid type, and their cost would make even a moneylender blush. Also, charging rates from the various generating sources must be optimised to get the best out of them – they prefer to be charged at higher voltages than lead-acid batteries. For most yachtsmen, the initial costs probably outweigh the benefits, but for the man looking for supreme reliability they deserve very serious consideration.

Capacity – How Much?

It is important to realise that stated capacities are for fully charged batteries. And the notion that you can count on getting, say, 100 amps out of a 100Ah battery is a rash delusion. To flatten a battery entirely is to risk serious damage, so you don't want to wring the last drop out of it anyway. Good quality deep cycle batteries will tolerate regular discharge to, perhaps, fifty per cent of total capacity without undue harm, but could baulk at being asked to repeatedly do much more than that (though the occasional discharge to lower levels should present no serious problems). Secondly, the standard charging facilities you have at your disposal are likely to be fairly rudimentary – incapable of ever bringing the batteries to full charge, or even near it. Batteries charge rapidly to about eighty per cent capacity, after which the charge rate is tapered off. So, sandwiched between the probable maximum and safe minimum we have eighty per cent minus fifty per cent – a puny 30Ah residue from the original 100Ah capacity.

So, if we take our specimen sailboat 'at sea', with its 120Ah consumption over twenty-four hours, we should obviously need four such batteries (400Ah), to run just for a single day without charging. This once seemed excessive for a mid-size yacht, but is now commonplace. But many of us still get by with less by managing our electrical consumption intelligently and, of course, by providing support charging at regular intervals. My own boat in offshore mode would consume no more than 40Ah per day in similar circumstances, and we operate her successfully with a total of 240Ah of battery capacity, supplied by two of the best quality batteries we could buy. But we can only scrape by with this by making electricity as we go along. Which leads us neatly to that whole intriguing subject . . .

Feeding the Beasts

I was once approached by a man clutching brochures. He entered my office and waved them about a bit before saying: 'I'm thinking of fitting an autopilot – and my boat is already bursting with electrical gizmos. Do you think I need larger batteries?'

His question highlighted a popular misconception. In the minds of many, more demand simply signals more battery capacity. But this is as illogical as attempting to increase your bank balance by ordering another cheque book. Although having a generous storage capability is unquestionably beneficial, what, of course, he actually needed was more electricity not a bigger box to put it in. The electrical appetite of the modern sailboat can only be satisfied by feeding it with the stuff it thrives on. And you have to plan to get it from somewhere.

Most sailboats rely solely upon shore power and engine charging to keep their systems running. If you spend enough time alongside or motoring, this could be sufficient, but the long-distance sailor has to be more resourceful. Let's consider the options one by one.

Engine Charging

When the aforementioned gentleman and I started to draw up an electrical appraisal of his boat we almost immediately came to the matter of his alternator. I wanted to know how big it was.

He looked baffled but, after some reflection, held out his hands about nine inches apart. 'And it's painted green,' he added by way of further elucidation.

Gently, acknowledging the ambiguity of my question, I explained that it was its output that was important, not its physical dimensions. He had no idea, he told me. It came with the engine, fitted to the boat when be bought it. He assumed all alternators to be the same.

And he is by no means alone. Most of us think that engine manufacturers know how many onions make five and can therefore be relied upon to fit the appropriate bits of kit to their engines. So long as it's painted green, or red, or grey, or whatever colour matches the rest of the engine, we believe we can safely regard it as an integrated bit of the whole, adequate in every way.

Not so. When I bought the nearly-new demonstration engine I fitted to *Spook* I was disappointed to find that the alternator had an output of only thirty-five amps – nothing like enough for the plans I had for it. I then discovered that a fifty-five amp replacement from the same engine manufacturers would entail a further cost equal to about twenty per cent of the list price of a whole brand new engine – a quite ridiculous surcharge on something which should have been a standard item in the first place. Of course, marine engine manufacturers exist in a fiercely competitive market, but there is no excuse for supplying equipment not up to the job.

So, when planning overall charging strategy, the first task the owner must face is to establish how much he can expect from his engine. The maximum rate of charge a deep cycle battery (or a number of such batteries connected in parallel) will accept without suffering internal damage is about twenty-five per cent of total rated capacity. For example, a heavily discharged 200Ah battery may be charged at fifty amps. But, as alternators can overheat and burn out if asked to deliver their full rated output continuously, we also need some overage as a buffer. In this case perhaps a minimum of 65Ah would be ideal. And if you had, say, 400Ah of total battery capacity, you might be looking to fit a 120Ah alternator – with an important proviso: at full output, an alternator of that size will absorb about 2hp from the engine, so you must ensure you have that extra power to spare.

To get the best out of your engine, you should first make certain that alternator output matches battery capacity. Not to do so means that your engine will be squandering some of its potential generating capability.

Most marine engine charging systems have been handed down from automotive practice. This is unfortunate because the requirements are quite different. In just starting the engine, car batteries are rarely depleted below five per cent of their maximum capacity, whereupon they are quickly restored to full charge by the alternator. But, as we have already covered, yacht batteries are often heavily discharged, and require more than a quick top-up to bring them back to good health. The standard automobile type alternator and regulator is not well suited to bringing batteries back from deep discharge.

The regulator controls alternator output, typically at 14.0–14.2 volts, by monitoring battery voltage. The deeply discharged battery will, at first, be below the level of regulator control and will accept all that is thrown at it, dissipating any excess current as heat. However, as the battery charge level rises, the regulator senses this and progressively reduces the alternator output. Unfortunately, it will almost certainly have been deceived. For, the movement of electrically charged ions within the battery cells (which converts the lead sulphate to lead dioxide in the charging process) starts on the outside of the plates and works its way inwards by diffusion through the active material. The effect of charging on the outer surfaces of the plate will rapidly show full-charge characteristics and fool the regulator into thinking that the whole depth of the plate is charged. It may actually take many more hours of further charging (all at the measly output permitted by the stupid regulator) to efficiently recharge the battery. Think what that means in terms of fuel costs and engine wear, let alone the damage done to the batteries by only partial charging!

This damage, incidentally, is largely caused by sulphation – the killer disease of batteries. When lead sulphate is produced by battery discharge, it is initially in a soft spongy form which can be readily converted back to lead dioxide by recharging. If the battery is left in a discharged or partially discharged state for long, the lead sulphate hardens into crystals which will be much more resistant to reconversion.

The 'Intelligent' Controller

Recognising car-type regulators as rather sorry devices which should long ago have been consigned to some historic maritime dustbin, manufacturers have been active in developing more efficient alternatives. These take the form of charge controllers which either work in parallel with existing regulators or replace them altogether.

Typically, these gadgets take over control of the alternator and introduce a charging regime tailored around the specific requirements of deep-cycling batteries. A pattern of fluctuating voltages is imposed on the battery, alternately boosting and resting it, to allow full ion diffusion through the depths of the plates. This ensures maximum charge levels which, in turn, go a long way towards preventing sulphation. As well as being voltage sensing, the most sophisticated of these units also monitors the battery's ambient temperature. A microchip then calculates the existing battery voltage with regard to temperature and adjusts the alternator output to the theoretically most suitable charging value.

There is no doubt that anyone wanting to get the most electrical benefit from his engine should consider one of these microelectronic miracle machines, but a word of caution! Enhanced charging performance places strains on your primary power system. It is very important that your batteries and alternators are correctly matched and that all cables be of the correct size.

The 'Green' Machines – Wind Generators

Many people regard them as unsightly, but few long-distance sailors can get along without one of these thoroughly useful devices mounted somewhere on their boat. For those whose engine-driven generating capacity falls short of their needs, this is probably the first step to consider.

There are two distinct types of wind generator:

- Alternators. In these a permanently magnetised rotor spins inside a ring of coils known as the stator. As can be deduced by the description, an alternating current is produced and must be rectified by silicon diodes before being fed to the battery. All commonly available smaller units are of this type.

- Electric Motor Types – usually based on standard off-the-shelf units. In these the motor's original mode of operation is reversed. Instead of applying electricity to it to make it turn, the motor (now a generator) is spun by the wind, and generates an electric current. No diode is needed with this type, as rectification is accomplished mechanically by the motor's own commutator. The innards of the motor type generator differ from the alternator in that, rather than a magnet spinning inside a ring of fixed coils, here the coils are mounted on the rotor (now called the armature), and turn inside a ring of fixed magnets. Larger wind generators are more likely to be of this type.

The wind, of course, is a very variable energy source, and it will come as no surprise to learn that the harder it blows, the faster the generator will turn, and the greater will be its output. To some extent this is to be welcomed, but it is easy to have too much of a good thing. When the wind pipes up, all wind generators are capable of producing very high voltages which, if not controlled, will damage batteries and other electrical equipment – including the generators themselves.

Protection of the generators can be provided by mechanical governors which feather or stall the blades at a prescribed rotational speed, or by thermal cut-outs which switch the unit off before the stator coils reach harmful temperatures. A word of warning here: manufacturers' advertised outputs are sometimes quoted for 'cold windings' speed – conveniently ignoring the fact that the coil temperatures associated with that speed may switch the generator off altogether after a few seconds! Certain types of generators – those with iron-cored stator coils – are self limiting and cannot burn themselves out (though still can be mechanically destroyed by excessive speeds). And, as this type can do without a thermal cut-out, and will therefore operate continuously, the manufacturers claims are more likely to accord with real life performance. However, even this seemingly perfect apple has its worms: the type is characteristically heavier, and the magnetic drag effect of the iron-cored coils (sometimes described as cogging) makes them poorer performers in light conditions.

Whereas the enemy of the generator is self-produced heat, it is the surges in voltage that will wreak havoc with the batteries and other electrical appliances. Unless the generator is a very small one, or the battery capacity very large, it will be necessary to fit a voltage regulator to control the generator's most boisterous excesses. Preferably, the regulator should be of the type that monitors generator output rather than battery input. If that sounds like the same thing, remember that the battery is being fed from more than one source. For instance, an input sensing regulator would sense and reject the high voltage peaks from your expensive 'intelligent' engine alternator controller – rendering it useless. Figure D64 shows the correct circuit for a two battery installation.

Most wind generators are self-aligning, rather in the manner of a weathercock, but many ocean voyagers prefer to have the alignment under their control – 'farming' the wind by manually rotating the unit to give maximum output when it is needed, and feathering it when it is not.

Wind generators – particularly those with two or three long blades – can be noisy things. Many people find this disturbing, even unbearable, but personally I have no great objection to it. Indeed, in one sense their noise is reassuring. When in your bunk at night, the variations in pitch and volume give a valuable indication of the wind strength. On a number of occasions the wind generator has called me on deck to check the anchor – once with every justification.

Electrics

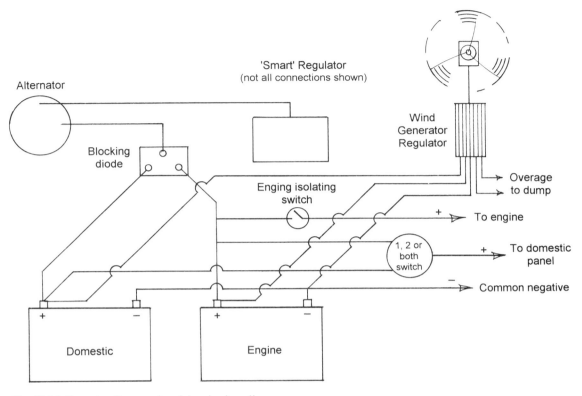

Fig D64 Two-battery electrical circuit

But beware. Their blades whirl at amazing speeds and can inflict serious injury. They must, if at all possible, be mounted high enough to be safe. On one boat I knew, they named the wind generator Geronimo because of all the scalps it had taken.

Solar Panels – the Silent Source

It is tempting to think that this might be the universal solution to all charging problems. There is something appealing – downright miraculous, even – about the notion that you can seemingly get something for nothing. And the advertising reinforces this. 'Free electricity from the sun!', it howls, and off we all trot to spend our money.

Unfortunately, this turns out to be quite a lot of money, if viewed from the cash per amp angle. Solar panels are expensive for what they provide – and far less efficient, viewed in those terms, compared with wind generators.

Each panel consists of an array of cells. The cells themselves are made up of slices of specially treated silicon crystals. A single cell will produce about 0.5V under good conditions, and a number (either 30, 33 or 36) must be wired in series to produce

The Innovative Yacht

A twin panel solar charger array which can be angled to face the sun.

a voltage sufficient to charge a 12V battery. Amperage is proportional to the area of your panel, and is gained by wiring seried cells in parallel. Obviously the bigger the panel, the more the output.

Output figures can be misleading and you should examine them carefully. In presenting their wares in the best possible light, many manufacturers quote 'average' outputs derived from a technically perfect installation operating in ideal climatic circumstances. In reality, conditions are usually less than ideal and you may end up disappointed in the actual performance you get.

Of course, it is easy to understand how solar panels would not work well in areas such as Britain, where the sunlight is weak and the clouds plentiful. But what about the tropics where the sun is rarely shrouded and hangs poised overhead to pour all that free energy down onto your panels? Sadly, here we have the most frustrating irony: for, as temperature rises, output drops. And solar cells can get very hot indeed. After all, they are black (or blue which is nearly as bad) and we deliberately mount them in the sunniest places. Quoted output is usually rated at 25°C (77°F). At double that temperature – easily achieved in sunny climes – the output could be seriously impaired. Panels having a larger series of cells (33 or 36) are less punished, but the performance loss is still significant. Probably the ideal location for a solar panel is a polar ice-cap in summer.

But we should not be too dismissive about solar panels. Although they are unlikely to make substantial contributions towards your energy requirements – unless you adopt an impractical and incredibly expensive wall-to-wall approach – they are excellent for keeping the batteries topped up when the vessel is unattended. One would be loathe to risk leaving a wind generator to its own devices during stormy winter months, but a solar panel will provide an entirely safe and very useful trickle of power to keep batteries alive and kicking.

Shore Powered Battery Chargers

Self-sufficiency is an admirable thing, but why turn down an easier alternative when it is available? Marinas are usually bursting with electricity, so you may as well make use of it while you are there.

One of the problems you come across when selecting equipment is comparing like with like. Rarely is this more the case than when choosing between battery chargers. You can wander into any discount motor accessory shop and find chargers which will cost you no more than a few pounds, whilst specialised marine equipment could knock you back hundreds. So what distinguishes one from another – the cheap and cheerful from the downright expensive?

Complexity and facility are the short answers. We should pay for what we need – no more, and certainly no less. The weekend sailor may be happy to have his batteries topped up slowly during the week, but a sailing school boat would want to bring depleted batteries back to life within hours. And, if you buy too cheap, serious battery damage could occur; buy more than you want and you will have wasted the money.

So, how do they work? Battery chargers are basically devices which convert alternating current (AC) mains voltage (240V in the UK and Europe, 115V in the USA) to somewhere near battery voltage, and then rectify the output to the direct current (DC) the batteries require for charging. If this was all there was to it, this could be accomplished by no more than a transformer and a handful of diodes, and this is exactly what you get in the very cheapest units.

At the bottom of the stack are ferro-resonant type chargers. These are very simple devices which have no regulating circuits, relying instead on the way the transformers are wound to control output – usually 13.8VDC for a 12V system. Ferro-resonant chargers are naturally responsive to battery voltage. If the battery voltage is low, the charger output rises (within its limits) in an attempt to maintain a constant level. Then, as the battery level rises, the charger output tapers off accordingly. For instance, with the battery only seventy-five per cent charged, the charger output will already have dropped to a mere trickle.

And therein lies the principal problem with this type. For, to maintain efficient charging, it is necessary to maintain a healthy voltage differential over battery voltage at all times – easy to do at first, but increasingly tough as the battery is progressively charged. Ferro-resonant chargers are inherently inflexible. The output voltage of 13.8VDC is a compromise struck to optimise charging (within the crude limits of the type) whilst protecting against overcharging. For, if high voltage differentials were maintained throughout the charging cycle without some external control, the risk of serious damage to the battery would be very great indeed.

Yet ferro-resonant chargers do have some advantages. As we have already covered, their cost is low, and they are also extremely reliable – a natural result of their simplicity. For the weekend sailor whose batteries could be conveniently trickle charged, they might be the best overall choice.

But for greater efficiency we need more sophisticated control. Occupying the middle to upper ground of the charger hierarchy are what are known as triac and SCR (Silicon Controlled Rectifiers) chargers – the difference between the two being that one exerts its control on the power side and the other on the battery side of the transformer. These types seek to regulate the output in such a way that the voltage is maintained over a broader band of the charging cycle, whilst still guarding against overcharge. In terms of sophistication, there is considerable spread in this group. The most basic have very simple voltage controllers, whilst the top-of-the-line models have 'smart' microprocessors which impose a charging program optimised for each application.

Until recently, triac and SCR chargers represented the pinnacle of charger technology, but they have now been displaced by an ingenious new type known as switch mode chargers or, more colloquially as switchers – a word which describes them admirably. A switcher charger works by taking in mains voltage (240VAC at 50Hz frequency in the UK, 115VAC at 60Hz in the US) and converting it to DC at the same voltage. Then, by rapidly switching it on and off, it converts it back to AC but at a greatly increased frequency – as much as 100,000Hz in some units, and they will probably do better as technology advances. This is then dropped to battery voltage and rectified back to DC, the whole operation being controlled by microprocessors.

This is clearly a very complicated way of producing what is, after all, a rather modest DC voltage, so there have to be some important gains. And there are. For a start, the control is very precise, with the voltage optimised for every stage of battery charge. Switchers are also small and light – typically a quarter of the size and a tenth of the weight of a comparable triac or SCR charger. And they operate without generating nearly as much heat – all of which is wasted energy in conventional rigs.

It is probable that switchers will dominate the charger market in years to come. Although relatively pricey now, their initial cost should be measured against their efficiency and the consequential extension of battery life. And, as more charger manufacturers move in this direction, prices will probably come down.

Before we abandon this subject, it is worth reflecting on the declared output for the various types. Some claims verge on the fraudulent and, when making comparisons, it is easy to be misled. If a charger claims to be a high-output unit but can only maintain the declared rate of charge for a small part of the charging period, then plainly you are being deceived. It is the overall performance of the equipment that matters – not a momentary peak. Be particularly suspicious of the cheaper ferro-resonant chargers which, by their very nature, start with a flourish and end poorly.

And let me repeat what I said at the beginning of this chapter. Electrical systems must be viewed as a whole – not piecemeal. Everything must be balanced and matched: batteries, charging devices, regulators, and the very wiring itself must be properly sized to meet the demand, and to live comfortably with every other component in the system. Get it wrong and the results can be disappointing – even disastrous. Get it right and your electrics will be transformed.

Despite all the electronic marvels of our age, there is still plenty of use for traditional lamps.

Chapter 13
Electronics

I started this chapter in the certain knowledge that events would overtake it. Developments in electronic navigational aids are not so much advancing as hurtling along. A reflective glance astern tells us how far we have come, and in how short a time. It was only in about 1980 that satellite navigation became available for yachtsmen; and, since then, we have already discarded one viable system (Satnav) and put another (the Global Positioning System – GPS) in its place. Twenty years earlier than that, Decca and Loran receivers were about the size of juke boxes and so expensive that even many merchant ships did not carry them. For the sixties yachtsman, such toys were quite out of reach. About all he could hope for was a log, an echo sounder, and a radio receiver with RDF facility – all rather crude, unreliable, and also expensive by the standards of the day.

And, if we wonder at how far we have progressed, then how much further have we yet to go? At first glance it would seem that we have got everything pretty much covered in the gizmotic sense: Interactive instruments, minuscule radars, weather faxes, GPS navigational systems, chart plotters, autopilots, and so on and so on. But to think we might have seen all, would be to underestimate the ingenuity of the technological wunderkinder and, of course, the commercial incentives that inspire them. For marine electronics is very big business indeed. Although other sections of the industry have slowed and even withered, our appetite for digitised delights has grown a dozen-fold.

And who are we to complain? After all, navigation is essentially about information. In order to make seamanlike decisions, we need to know where we are, where we are bound, and what is going on around us. In the past this was easier said than done. Dead reckoning was logical but needed care and judgement; and celestial navigation was arcane enough to prevent all but the most experienced from venturing too far from land – perhaps no bad thing, some would say. But although some old salts might still hanker for the heft of a sextant and the soup-stained pages of their sight reduction tables, electronic aids are here to stay, and there is no reason why we should not make good use of them, provided we have the back-ups to hand should they fail.

This last point is important. The man who relies entirely on electronics is taking a big risk. In the late seventies I was virtually becalmed in the middle of the Gulf of Mexico when a large power-boat hove into view and altered course towards us. Panic ensued! The spirit of Captain Morgan still presides over that area. Piracy is not unknown. Nervously, we broke out the boat's armaments and, with less heroism than we had anticipated for such a situation, lay them out of sight behind the cockpit coaming. The power-boat circled us, some fifty yards distance, and a man with a hand-held bullhorn emerged from the wheelhouse.

'Can you give me course and distance to Biloxi?' he hailed. 'The Loran's crapped out and I'm lost.' A few minutes later, after thoughtfully buoying-off a very welcome six-pack of cold beers for our delectation, he was on his way back home to Mississippi.

It is prudent to realise that, wondrous though they are, electronic navigation aids are exactly that – aids. They can never be substitutes for traditional skills or experience. Remember Murphy's Law ('if it can go wrong, it will') hang on to your charts and sextant, and know how to use them.

It is very easy to be seduced by electronic gadgetry. It flashes and bleeps before our senses with more allure than a siren's call. Many of us succumb to temptation. In my work as a marine surveyor I find it is depressingly common to see yachts expensively burdened with a magnificent array of navigational instruments, yet deficient in such mundane items as anchors and storm canvas.

Of course, if money is no object you can buy the lot, but this is a condition unknown to many of us. For those with shallower pockets, it may be advisable to assess our needs rather than our wants, and to plan our purchases accordingly. A fairly minimalist approach for different types of sailboat might generate lists looking something like this:

- Cruising Only: Navigator (Decca, Loran-C, GPS) and depth-sounder. The latter should be visible in the cockpit, either by placing the instrument there or by fitting a repeater. If planning a lot of inshore cruising in crowded or poor visibility areas, then a radar would also be handy – though by no means necessary.

- Cruiser/Racer: Navigator, depth-sounder and electronic log. Possibly a radar, though the weight and windage aloft might put some skippers off.

- Racer/Cruiser: Navigator, depth-sounder, log, and wind speed and direction indicators, with repeaters as appropriate. A radar would almost certainly be rejected, but a chart plotter might be useful for the navigator in a hurry.

- Racing Only: As above, but with expanded close-hauled display on the wind instruments, and such additional goodies as Velocity Made Good (VMG) meters.

Large display cockpit repeaters are popular. These allow small twitches in wind and speed to be easily monitored in the heat of battle.

- Ocean Racing or Cruising: All of the other appropriate gear plus a weatherfax. The singlehanded sailor might also require an electronic compass for its off course alarm facility.

Inevitably, these lists could be adapted and expanded to suit individual requirements. Some very useful gear – EPIRBs and Navtex for instance – do not even appear, and neither do communications devices such as VHF or SSB radios. But many people buy navigational aids in the wrong order, and their choice is often dictated by the relative fun they hope to get from each bit of kit. Logs are undoubtedly more entertaining than depth-sounders – 'ten knots on the clock and we hadn't even popped the kite' – though they do nothing to keep you out of danger. VHF radios can give hours of endless amusement – 'Bold Buckaroo to Sailbad The Sinner, do you read me?' – in the transmission of trivial information, and are often the first things put aboard, even before a compass.

Another way to look at it is to ascribe relative values to each item. Taking absolute necessity as being 100 per cent, every sailor should be able to rationally assign comparative values to each piece of equipment. My own views on the subject are listed below. Not all of the items rightly belong in this chapter but needs vary with their alternatives:

Magnetic compass – 100%
Depth-sounder – 100%
Sextant, chronometer and tables – 100%
Trailing mechanical log – 90%
Navigator (my choice would be GPS) – 85%
VHF radio – 80%
Autopilot – 75%
Weatherfax or Navtex – 60%
Electronic log – 50%
Wind indicator – 40%
Electronic compass (assuming autopilot has its own) – 30%
SSB radio – 25%
Chart plotter – 25%

Interface or Stand-Alone

When electronic aids first emerged they were totally independent units, dedicated to the task for which they were designed: Logs measured speed and distance run, depth-sounders told you how deep the water was, and wind indicators were expensive weathercocks – with none trampling on another's territory. Although manufacturers

Electronics

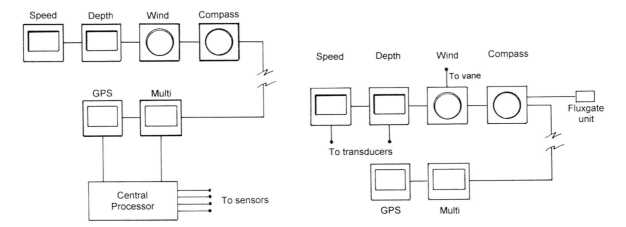

Fig D65 Central processor linked navigational instruments

Fig D66 Daisy-chained navigational instruments

produced matching suites of instruments, the similarities were cosmetic; they were operationally unconnected and yachtsmen could take their pick between brands. To some extent this is still true, but not always.

The name of today's game is integration. This is partly clever marketing – an encouragement to customer loyalty – but it does also simplify the wiring. Manufacturers usually adopt one of two systems, namely:

1. A group of instruments centred around a remotely located central processor. All transducers and sensors are led to this 'black box', which processes the information and passes it on to the various display units (Fig D65). Advantages of this approach are that the processor can be placed somewhere protected and dry (making the whole system less vulnerable to damage) and also that the routing of the relatively thick co-axial transducer cables can be made easier and neater – usually hidden behind the yacht's furniture. On the debit side it can be argued that, although the central processor might be safer, a serious malfunction in it could knock out all of the instruments at a stroke.

2. The more common method is to link the various instruments together like blooms on a daisy chain (Fig.D66). Each instrument has all of its own logic circuits contained within its own casing, but the interconnecting wires supply the power and allow the exchange of information between each one. Every instrument will work independently and you can add new units or remove defective ones without affecting the others. With this approach each instrument is perhaps more

susceptible to damage than a centralised system, but the instrumentation as a whole is less vulnerable.

In practice there is probably not a lot to choose between either philosophy. Modern electronics are very reliable and, unless actually physically destroyed or invaded by water, tend to run without problems for years. For those planning to install a comprehensive instrument package, either approach would be fine. But the sailor who wants perhaps just one or two instruments, with the option perhaps to add others later, then the daisy chain method makes more sense.

For their internal communications, integrated systems usually use their own information 'language' – though the various buy-outs and amalgamations within the industry have muddied the water in this regard. Although every manufacturer would prefer thus to trap the consumer into an exclusive commitment to his products, this has proved impracticable. Luckily, there is a universal protocol that nearly every instrument understands (though some not as well as others). This is known as NMEA 0183, in which the initials stand for the National Marine Electronics Association – the 'national' in that title meaning the US – and the numbers for a series of such protocols which started with 0180. When interfaced with more modern kit, devices designed around the earlier NMEA numbers (particularly navigators and auto-pilots), may not work fully in the manner originally intended.

So, having opted for one or other of the system architectures, there is now the matter of the instruments themselves. In technical excellence and constructional quality, there is not an enormous distinction between most of the leading manufacturers, so the decision boils down to whichever offers you the kind of facilities you want, tempered, of course, with your personal taste. Rather than reviewing the myriad 'special' features on offer – most duplicated by more than one manufacturer – I think it would be wiser to deal with the subject in general terms.

If 'basic' is a word we should use when talking about electronics, then that might be a good place to start:

Depth-Sounders

Basic, perhaps, but in my view the most essential. Depth-sounders tend to be regarded as the Cinderella instrument – nice to have slogging away uncomplainingly but of no entertainment value whatsoever. Few people get excited about them.

And yet they are a potent navigational aid that can get you safely home when other systems have failed. I remember bringing a trimaran down the North Sea coast from Berwick-upon-Tweed, some years before Decca was generally available. A high pressure system hung over the region. The wind was light. Visibility was less than a quarter of a mile, and the RDF stations – one almost due North and the other due South – were useless to me. But, once clear of the rocky hazards of the Farne Islands, we made it down to Blyth (and the convivial comforts of the yacht club there) simply

Electronics

Cetrek C-Net instruments.
A typical example of a modern integrated instrument system.

by short tacking along the ten-fathom mark – not an extraordinary ploy at all then, but the sort of trick which push-button navigation might encourage us to forget.

By modern standards the sounder I had then was a puny thing. Its two scales offered ranges to sixty feet or sixty fathoms, but the latter verged on wishful thinking. By contrast, the Stowe Navsounder we now have fitted to *Spook* returns reliable soundings anywhere on the continental shelf and we use it constantly when in coastal waters.

A useful 'extra' in most depth-sounders is the anchor watch facility. The sounder registers any rapid rates of change in depth (gradual changes as with the rise and fall of the tide are tolerated) and a warning buzzer alerts the crew. Some depth-sounders also provide shallow water and deep water alarms, the limits of which can be set by the user.

Displays can be either digital or analogue, with still a number of the rotating, flashing diode types being popular. Though comparatively expensive, the latest developments include forward and side scanning depth-sounders which display a graphic representation of the underwater profile on an LCD screen. One claims to see ahead up to five times the depth over a flat bottom and up to nine times the depth over a shelving bottom. Yet more, that a hard vertical object, like an isolated rock or a wreck, can be seen at over sixty metres. Its not hard to imagine how useful that could be when entering a strange anchorage.

Electronic Logs

These measure speed and distance travelled through the water. Although there have been flirtations with Doppler-type sensors (which have no moving parts and create no extra drag), the vast majority of logs obtain their input by magnetically counting the rotations of small spinners or paddle-wheels mounted below the waterline.

Extra functions include such bonuses as trip meters, timers and (for the racing man) count-down timers for those frenetic minutes before the start.

Total reliance on logs, whether electronic or mechanical, has led to the downfall of many a dead reckoner – unfortunately with 'dead' sometimes being the operative word. The problems arise with fouling of the sensors. If the paddle-wheel or spinner picks up any weed or other debris, or simply silts up through disuse, then it will rotate slower (if at all) and will under-read. This can fool the sailor into believing that he has travelled a lesser distance than he actually has, which can come as a nasty surprise if he arrives early at some navigational hazard such as a shoal or reef.

As with other electronic instruments, faith in their accuracy is reinforced by the certainty with which the data is displayed. Ask a novice how fast the boat is going and, after a glance at the log, he might tell you 'Five-point-four-three knots' – duly impressed with the two decimal places. Of course this is nonsense. Even if properly calibrated (which is rare) there are too many circumstantial variables for any instrument to be correct within a hundredth of a knot. Electronic instruments are very useful tools, but we should guard against being too overawed by them.

Both analogue and digital displays can be had, with the latter probably being the most popular. Analogue instruments usually display distance run digitally on small inset LCD windows.

Electronics

Wind Indicators

The Shetland islanders know a thing or two about the wind because they get a lot of it up there. Tied up alongside in Lerwick, with some energetic carousing going on in our cockpit, I once caused derision amongst a group of visiting locals by switching on the anemometer to see how hard it was blowing. 'T'would blow the whiskers off a cat, laddie', I was told, and the bottle got passed around again.

Over the span of time that has elapsed since then, I think I have learned to agree with them. The wind, in both its speed and direction, is probably the most discernible of nature's phenomena, so why should we need to measure it? Tell-tales on the sails, and a wisp of wool tied to the cap shrouds will show you all you need to know – in daylight at least. The sea itself will tell you of any ferocity. For the cruising sailor, the wind indicator in its simplest form is one of the most dispensable instruments in the line up.

But, when interlinked with GPS, Decca or Loran, it can become a VMG (velocity made good) indicator, which is as handy a bit of kit for cruising as it is for racing. To judge which is the 'making' tack when beating, is one of the most recurrently crucial tasks that face any navigator. And, when running downwind, there is often a delicate trade-off between just sliding down the rhumb line towards your destination, or hardening her up for a bit more speed. Such decisions are easier with a VMG capability.

Although some LCD displays exist, these usually present wind direction in some representation of analogue format. Most manufacturers still favour the moving needle approach, which is clearer to the observer. Wind speed is usually displayed digitally.

Electronic (Fluxgate) Compasses

Such compasses rely upon a sensor which senses the earth's magnetic field with an arrangement of coils. Unlike the traditional compass, there are no moving parts. The fluxgate can be located remotely, out of harm's way and at a distance from large ferrous objects such as the engine.

Of course, being essentially magnetic devices, fluxgates are subject to the same variations from True North as traditional compasses. However, the errors induced by deviation – the induced effects of the yacht's own magnetic characteristics – can be eliminated by the electronic compass's own software after following a simple procedure – usually just swinging the yacht through 360°. Fluxgates are especially useful on steel boats, where the problems with deviation can be immense. In these cases, the sensor can be mounted high on the mast, away from the strongest influences.

Apart from the steering compass display itself – which can be either digital, analogue or both – an important role of the electronic compass is to provide heading information to other instruments within an integrated system. Autopilots are totally reliant upon fluxgates, and so are those that attempt to predict events – some chart plotters and radars. For the single-handed sailor, the off course alarm facility would be extremely useful.

Position Fixing

This lies at the heart of all navigation. Clinging to the coast is straightforward enough, but the open oceans are sadly lacking in signposts. Dead reckoning will guide you well enough over short distances, but on longer passages the accumulation of errors progressively devalues such estimations to the point of guess-work.

Early navigators were quick to find that they could determine their latitude by observing the altitude of the sun at its highest point in the sky – a 'noon' sight – but the calculation of longitude eluded them until the invention of the chronometer by an Englishman, John Harrison in the 18th Century (for which he was awarded a Government grant of £20,000 – a truly immense amount of money by the standards of the day). That event was absolutely pivotal in the history of exploration. Navigation was then, and still is, totally dependent upon the measurement of time. Given an accurate clock, man can roam at will.

Although dead reckoning and celestial navigation still have their place – a primary place, many would say – in practice their day-to-day usage will inevitably give ground to electronic navigation, as use of the lead line has already yielded to the depth-sounder.

These electronic navigation systems fall into two groups: Those reliant upon land-based transmitters (RDF, Decca, Loran-C) and those that use satellites (Satnav, GPS). With transmitters gradually closing down, RDF is becoming less and less useful and, for the purpose of this chapter, I intend to ignore it.

Historically, we have to thank the Second World War for both Decca and Loran (then called Loran-A), which were developed independently for use by the Allied forces. Although there are similarities between the systems – they both use groups or 'chains' of transmitters working in concert – there are also essential differences.

Loran-C
This system operates with master and slave stations transmitting precisely timed pulse envelopes, from which a coarse position is derived by measuring the time it takes for each pulse to arrive from each station in the group – the further away you are, the longer the signal takes to get to you. Fine adjustment of the position is made by comparing the radio frequency phases within the pulse envelope. There is good Loran-C coverage in North America, northern British waters, the North Pacific, Saudi Arabia, and much of the Mediterranean. Coverage will shortly be extended to cover southern Britain and much of France – hopefully by late 1995. There are no Loran-C stations in the southern hemisphere. With the exception of the inaccurate and little used Omega system (which I have also chosen to ignore) Loran-C offers the

greatest range, effective to about 1,000nm, and to even double that (though less accurately) with the sky wave.

Decca

This basically northern European system uses a continuous wave signal which is kept synchronously in phase between the master transmitter and its slaves. The receiver monitors the master transmitter and compares how far it is in or out of phase with its slaves. This produces hyperbolic position lines whose intersection determines your position. Older Decca sets required special latticed charts, but modern units simply give you the latitude and longitude in a continuously updated digital readout.

Decca has a maximum effective range of about 400nm, reducing to about 250nm at night. Within reasonable reception areas, accuracy is superior to Loran-C – better than fifty metres within fifty nautical miles of a chain – and the repeatability of a position is excellent. Take the same Decca installation on the same vessel to the same spot and the position shown will be identical – a facility much favoured by divers and fishermen, who often have need to find unmarked locations offshore.

There is an important difference between Decca sets carried on commercial vessels and their smaller (and cheaper) counterparts on yachts. Professional receivers continuously monitor all three position lines passing through the vessel on indicators called Decometers. Once set up, the position can be obtained without further action.

Smaller receivers monitor the three lanes sequentially, updating the position about every twenty seconds. And positions are not unique. Because in any continuous wave transmission the same relative point on the phase cycle repeats every 360°, most yacht sets must be initialised by first entering a coarse position, accurate to within three nautical miles – which establishes it safely within a single wave length. After that, the equipment will fine tune the position to within its capability and keep continuous track of the correct phase cycle in which it should reside. But if you were to switch the set off and move it to beyond the boundary of that phase cycle, the set would wake up lost and would deliver a false position. This is sometimes known as 'skipping a lane', a deceptively merry phrase to describe a potentially dangerous event.

I once accidentally fell into this trap myself. Anchored in Studland Bay for the night, I slipped out just after dawn, forgetting at first to resuscitate my Decca. Shortly afterwards I remembered and, without bothering to check, assumed I was still within the correctable envelope. I was wrong, and lucky not to pay a heavy penalty for my carelessness; for eight hours later we closed the French shore over four miles adrift of the displayed position. In bad visibility, our trip might have ended disastrously.

So what is the future of these land-based systems? With the establishment of GPS, they now seem redundant. Not so. As GPS is first and foremost a military facility controlled by the US Department of Defense, it could be made instantly inaccessible to all other parties should US national security demand it. This vulnerability to unilateral whim makes other nations understandably nervous of abandoning their own well-proven systems.

But there remains some competition between Decca and Loran-C, with the latter clearly gaining the upper hand. Currently, Loran-C is operated by the US Coast Guard in American waters and by the US Department of Defense elsewhere. With GPS up and running, the US government sees no reason to continue funding its overseas maintenance of Loran-C, and is handing over its existing stations to the host countries. Denmark, Norway, France, Germany, Ireland, and The Netherlands agreed to set up a Loran-C system (called the North West European Loran-C System – NELS), but the British government – perhaps in patriotic defence of home-grown Decca – opted out and now have no active part in the decision making.

The upshot of all this is that the long-term future looks healthy for Loran-C but bleaker for Decca. With its greater range Loran-C can cover a larger area per unit of money spent. In its role as a locally controlled back-up for GPS, its lesser accuracy is outweighed by its operational cost effectiveness. It is likely that Loran-C coverage will flourish.

Which brings us to the bright new stars of electronic navigation – the satellite systems. The first to become available to yachtsmen (in the early 80s) was developed for the US Navy and was called Transit or SatNav. This consisted of five satellites circling the earth on roughly polar orbits at an altitude of about 1,075km, with the earth rotating within the 'bird-cage' form defined by the orbits. Receivers at sea level made use of the Doppler effect – the received frequency decreases as the satellite passes overhead – to compute a geographical position. Only a single satellite is necessary to obtain a fix. Although accuracy is good, the wide spacing of the satellites, meant that fixes could be as much as three hours apart at lower latitudes. The limitations of SatNav have spelt its end. Although still maintained in operation, as the satellites decay they will not be replaced and eventually the system will simply disappear from use. Indeed, it may already be defunct by the time you read this book.

Global Positioning System (GPS)

Here we have what is already proving to be the principal navigational facility for both professional and amateur mariners alike. Developed in the US for military purposes, a degraded version is available for civilian use.

GPS uses a total of twenty-four satellites (twenty-one in use at any time with three 'spares') which orbit around the earth at 20,000km altitude, forming a grid-like cage with each orbit inclined at 55° to the equator. At any one time there are at least four satellites 'in view' anywhere on earth. As each satellite proceeds upon its lofty way, it broadcasts its exact position at very accurately predetermined times. The navigator's receiver records the time taken for the transmission to reach it and computes a position line (actually a position circle) which when crossed with other position lines gives a three-dimensional fix. The whole procedure is extremely time dependent

Electronics

Garmin GPS.
The graphically represented 'way ahead' gives immediate visual indication of cross track error.

and this is achieved by fitting each satellite with an atomic clock accurate to within a single second every 300,000 years!

In its unadulterated form, GPS is very accurate indeed – probably to within less than a dozen metres – but, because such accuracy has implications regarding the guidance of incoming hostile missiles, this is deliberately blurred by what is known as selective availability (SA) which reduces the precision to 100m or so – which is still pretty respectable. Although fine for most activities, there are some applications – surveying, racing around the cans maybe – where greater accuracy is required. For this we have Differential GPS (DGPS), an ingenious dodge which requires a shore-based monitoring station in addition to the satellites. Basically, the idea is simple: the shore station compares the slightly erroneous SA position from the satellites with its own precise geographical location. It then transmits appropriate corrections to all other DGPS capable receivers in the area. DGPS coverage is fairly scanty to date, but is increasing steadily. In most areas the service is free, but British yachtsmen must pay an annual fee for it.

The expansion of DGPS raises an interesting political point: If selective availability can be corrected, why bother to have it at all? Perhaps at some future date, the Americans will admit defeat and throw open undegraded GPS to everyone.

Yet despite its limitations, for all practical purposes the standard GPS, as we have it now, works very well – except perhaps in one regard. In common with most Decca and Loran-C receivers, as an extra facility GPS will give you your course over the ground (COG) and speed over the ground (SOG) at any moment, which it calculates in much the same manner as you would in your chartwork by applying time to distance travelled. Now, although the degraded precision imposed by SA might have little significant effect over longer periods, it can be grossly misleading if sampled just briefly. I was on a yacht recently which was doing three knots whilst blocked off ashore – the inaccuracy being introduced by variations in the SA position. Of course, the longer the period sampled, the greater the accuracy; and most sets either automatically alter the averaging period to suit different circumstances (the slower the speed, the longer the period), or allow the navigator to decide. Periods between one and 120 seconds are typical. And accuracy improves significantly once the boat is moving. The yacht's progress, no matter how lethargic, establishes a more coherent pattern which allows the instrument's software to make a better judgement of speed – usually to within a fraction of a knot. In this respect some sets are abler than others. Those that have the capability to simultaneously 'listen' to the greatest number of satellites clearly have more information to go on and can take a better stab at guessing what is actually going on.

Can the SOG/COG facility on GPS (or, indeed Decca or Loran-C) be a substitute for an electronic log? Can you dispense with that extra hole in the hull and that troublesome little paddle-wheel? I believe so, and have adopted exactly this approach on *Spook* (though keeping a Walker mechanical trailing log aboard for emergencies). Racing skippers might need constant feedback on boat speed, but the cruising sailor can manage very well without. Whatever inaccuracies might exist in any electronic navigation system, to have SOG instead of speed through the water is both valuable in itself, and a guard against the inevitable DR errors introduced by attempts to approximate set and drift. Both systems can deceive you a little, but the electronic log will deceive you the most.

Other Features

As well as SOG/COG and position fixing, electronic navigators of all ilks are capable of other tricks. These include:

Waypoints (WPT)

These are critical points along your course – usually heading changes or route terminations – through which you want to pass. On most sets, these can be entered singly (by punching in lat. and long. or, in some cases, range and bearing) or arranged sequentially to form entire routes, which are often reversible. Waypoint capacity varies, but 50–300 is typical. There are various methods available to arrange and label them. On request, the navigator will calculate the course and distance to steer to each waypoint or, via NMEA interface, will instruct an electronic autopilot to take you there.

Some sets also give the time to go (TTG), based on the SOG being achieved. An extension of this function is an estimated time of arrival (ETA) to the end of your route. An audible alarm gives warning of approach to each waypoint.

Cross Track Error (XTE)
This is the deviation from the course required to close a waypoint or other destination. It can be expressed numerically or, as is increasingly the case, in graphic form. Its position-finding function aside, electronic navigators are basically dead reckoning computers which can only deal with the information they have on hand. Tidal drift, leeway, and imprecise helmsmanship are not normally within their ken, but the collective results of these will show up clearly in XTE. This is a very useful function in cockpit mounted units, where the helmsman can see instantly what corrections are necessary to close the waypoint.

Man Overboard (MOB)
A valuable safety feature to help you in this most scary of all emergencies. Ideally, the MOB control should be a dedicated button which you can press instantly should a crew member go over the side. It enters your current position (like a waypoint) and will subsequently display course and distance to steer to get back to the point where the MOB occurred.

Event Markers
Locations other than waypoints can be entered to provide useful information. For instance, you might enter the position of a hazard you wish to avoid. Thereafter, your GPS set will monitor your position relative to it and could provide you with a bearing and distance off as required. The position of event markers and waypoints is often graphically represented on a secondary display which you can call up at will.

Anchor Watch
You can sleep easier with this little function engaged. The navigator is left active and, if the boat moves outside a predetermined zone, an alarm sounds. The response is not always instantaneous. Delays are built in to iron out spurious signal anomalies – but at least you should be awake before you drag too far.

Chart Plotters
My first introduction to these things was worrying. Retained to assist in the negotiations to purchase a large motor yacht in France, I embarked on the buyer's then present boat some time before a winter's dawn in Poole. We set off into a soggy southwesterly, with nothing very reassuring in the way of visibility. From the wheelhouse I could have seen almost nothing forward – even without the mist and spray. The bow was high, and festooned with fenders, neatly basketed in racks in our line of

sight. A profusion of instrument lights reflected from the windscreen, further impeding our view.

The owner snuggled back comfortably as we belted south at 25 knots. Arrayed before him were the radar and a chart plotter – the first I had seen in action. The autopilot was at the helm, the coffee pot was on. Never once did he attempt to look outside.

And that was the way it all went. From Poole to St Malo, borne along in a plastic projectile which might just as well have had the windows painted over for all the use they were for watchkeeping. We crossed three shipping lanes, scorched down the Alderney Race, and a few hours later – he nonchalant, I white knuckled – we were alongside and I was resisting the temptation to embrace the first Frenchman I saw. As an example of seamanship it was abysmal, but as an exploit which owed everything to someone else's brains it was marvellous.

Chart plotters display electronic representations of charts upon a small cathode-ray tube type screen. They are supported by electronic navigators – Loran-C, Decca, but increasingly GPS – whose generated position appears as a small symbol on the chart. As the yacht moves, so does the symbol – a gull's eye view of your progress across the sea. The information each cartridge contains is astonishing. The user can zoom in to a resolution of only 0.6m per pixel – detailed enough to show an individual marina pontoon on a chart of the English Channel.

The charts themselves are contained in small pre-programmed cartridges (the C-Map being the ubiquitous industry standard), which are updatable (for a fee) on a periodic basis by the manufacturer. As an option, future developments will see a yacht's electronic chart portfolio updated remotely (for a larger fee) via a satellite communication link and a computer modem. Imagine! A navigation mark might change, your chart plotter will know about it, and you will be oblivious that anything clever has ever happened.

In common with conventional charts, you can indulge in 'what-if' planning of your routes by pre-plotting your intended path and examining the options. Waypoints can be entered and, in conjunction with your planned speed over the ground, ETAs can be calculated. If linked to an autopilot, a simple instruction will then take you there, passing through each waypoint sequentially until you arrive at your destination – push-button navigation, quite literally!

Many chart plotters log the yacht's movements at predetermined intervals – usually thirty minutes – which can be printed out (via an RS232 interface) if you have the appropriate gear. Any other performance parameters – speed, depth, wind speed – which might also be fed in will be similarly logged.

The development potential for chart plotters is huge, and there is no doubt we shall marvel increasingly at them as new models emerge. They aspire to be far more than simply electronic substitutes for paper charts. The aim is to have them as totally centralised navigational centres on which all navigational functions will be executed and monitored.

If this makes you uneasy, you are not alone. Most experienced yachtsmen feel nervous about reliance upon gadgets – and only a fool would allow himself to be totally dependent. But, there is no doubt that chart plotters will carve out a permanent place for themselves aboard many yachts – and there is no reason why they should not. The important thing is to understand their limitations and inherent fallibility. For the time being at least, the use of official 'proper' charts continues to be required by law. And long may that be so.

Radar

One purist I know describes them as 'revolving restaurants'. More properly, the word 'radar' is an acronym for RAdio Direction And Ranging. Radars were exclusively big-ship gear until miniaturisation came along, but they are now common on even fairly small sailboats – and almost universal on large ones.

The basic principles are well understood: a radio signal is transmitted from a rotating antenna and bounces back from any solid objects in its path. The time taken for the signal to return gives range, and its azimuth (direction) can be deduced from the position of the antenna at the moment of transmission. The information is translated into a visual display, either on a cathode-ray tube (on larger sets) or liquid crystal screens on the compacter, yacht-type equipment. The 'picture', with all the displayed echoes, is refreshed with every rotation of the antenna.

As the heart of a centralised navigation system, radars compete strongly with chart plotters – though their roles are actually quite opposite. Chart plotters display what they have learned; radars report what they see. But already there is some merging of these two functions. On the more sophisticated radars it is possible to switch between radar and chart plotter modes, or to have both displayed simultaneously on a split screen. Imported information from other instruments – most usefully echo sounders and GPS – can also be displayed. This is known as multi-screening and is, perhaps, the direction that future developments will take us.

To review a few of the more advanced functions of the modern yacht radar:

- Guard Zone: Draw a circle around your yacht and an alarm will sound if any radar visible target invades it. The diameter of the circle is user defined.

- Watch (or Sleep) Mode: Radars consume electricity and you may not want them to run continuously. In Watch Mode, the antenna continues to turn but all other functions (transmitter, display etc.) become dormant. At specified intervals (5, 10, 15 or 20 minutes) the radar comes to life for a few seconds and looks to see if anything has entered the guard zone. If it has, an alarm sounds and the radar resumes continuous operation.

- Waypoints: Imports waypoint positions from GPS, Loran, or Decca and marks them on the display. Other functions such as TTG (Time To Go) can be shown.

The Innovative Yacht

- True Motion: In True Motion mode, stationary objects – land, buoys, etc. – stay still whilst yourself and all moving targets waltz around them. This makes it easier to track and differentiate between moving and non-moving targets.

- Off Centring: This extends your viewing area by allowing the screen to be offset.

- Automatic Radar Plotting Aid (ARPA): This is the collison-avoidance system used by commercial vessels, but now filtering down to smaller installations. ARPA plots an extrapolated extension of your own course and that of any other moving targets to see how the situation will develop if all current courses and speeds are maintained. From this can be derived CPA (closest point of approach) and TCPA (time to closest point of approach). Alarms can be set to advise you of any encroachment on pre-set CPA and TCPA limits. This facility is probably of more interest to high speed craft where things can happen quickly.

The lofty VHF antenna at *Fairlight's* masthead gives the maximum VHF range possible.

Chapter 14
Communications

The advances in communication equipment has matched that of all electronics. Today there is no technical reason (though I can think of a few personal ones) why anyone should be out of touch, wherever they might be.

Radio communications are strictly controlled by international agreement and most countries impose restrictions and licensing requirements upon users. In this chapter I have mainly indicated the British stance in regard to this. Different nations have adopted different procedures and people living elsewhere should check with their national authorities to obtain further details.

VHF

This is the simplest form of radio communication available to the yachtsman – and probably all that most will ever need. As well as boat-to-boat communications, it gives access to coastguard and other emergency services, and enables link calls to be made to shore telephones.

Also by international agreement is the allocation of the available channels, to one of six separate purposes: distress, safety, and calling; inter-ship; public correspondence; port operations; ship movement; and, in the UK alone, yacht safety. Other channels may also be approved locally for specific uses – for example, in Britain channels 80 and 37 (M) are reserved for communication between yachts and marinas.

Regrettably, the international accord stops at the United States, where, having taken their own counsel, they have their own channel frequencies and designated purposes – though some are common to both. Most VHF sets can be switched from the international to the US system, and vice-versa.

The range of VHF is roughly 'line of sight' – perhaps twenty-five miles or so at twenty-five watts output, though powerful shore-based stations can reach out further.

To maximise range, it is helpful to have the antenna as high as possible, preferably at the masthead.

VHF sets in the UK require a 'Ship's Licence' and also there must be an operator aboard holding a 'Certificate of Competence' – under whose supervision other crew members can operate the equipment. This certificate is easily obtained by completing a straightforward examination with very little technical content. In the United States there is no examination, and a certificate can be obtained by signing a declaration that you understand the regulations and can transmit spoken messages in the English language.

Single Side-Band (SSB)

This is a truly international communication system, working ship-to-ship and also ship-to-shore through coastal radio stations located around the world. Link calls can be made with telephone networks ashore and emergency (Mayday) transmissions can be made on 2182 kHz.

To work the marine SSB bands, the operator must hold a 'Restricted' licence and the boat must also carry a licence for the equipment carried, which must be of an approved type. The operator's licence is more demanding than that for VHF, but is still not technically difficult. Training courses are available which cover the required topics. Typically, these would take about three days to complete.

Medium Frequency (HF) SSB has a range of about 400 miles, but High Frequency (HF) SSB will reach out for thousands. As well as voice transmission and the reception of broadcast weather forecasts and news bulletins, SSB gives access to weather facsimile and other radio data services.

Amateur (Ham) Radio

For the long-distance sailor voyaging on a tight budget, this could be an attractive proposition. The ham community is an international brotherhood capable of communicating with the kind of nimbleness and speed which should be the envy of other more formal institutions. Through a ham I knew, I once obtained news of a friend in Mexico City, within hours of a serious earthquake there, which had virtually severed other means of contact. Groups of stations (individuals, actually) organise themselves into 'nets' which gather on the same frequency at pre-agreed times and exchange information. There are sailing nets in place world-wide, to which ham yachtsmen contribute with reports upon local conditions and other relevant gossip. For example, the UK maritime net gets together regularly on 14.303 mHz at 0800 and 1800 GMT.

Apart from a professional 'sparks' ticket, the ham radio licence is by far the most challenging to acquire. Basic electronic skills are required – very useful for the maintenance of all such equipment – and also Morse code to a speed of at least twelve words per minute. The study can be completed at home, but most people take advantage of organised courses on the subject – often 'evening classes' in the UK. Some people operate ham sets illegally, but are treated with justified distaste and little co-operation by legitimate users.

Amateur type transceivers are generally less expensive than SSB sets – and maybe a lot less if obtained second-hand.

Marine Telex

For the occasional writer on the move. Using a portable computer and a modem, any boat can be linked through an HF SSB transceiver to shore stations around the world. These shore stations run a 'mailbox' system, in which the subscriber contacts the station and retrieves any messages left there for him. Outgoing messages can be sent directly to any shore-based fax machine.

EPIRB

This rather ugly title is an acronym for Emergency Position Indicating Radio Beacon, which describes one of those pieces of equipment you hope you will never have to use. An EPIRB is a small self-powered waterproof transmitter that, when triggered, sends out a distress signal capable of being received by aircraft and satellites, and which then can be homed-in on by your rescuers.

There are two kinds of EPIRB: The simplest and least expensive kind transmit on the aircrafts' distress frequencies (121.5mHz and 243.0mHz). More sophisticated (and much more expensive) are those that take full advantage of the Global Maritime Distress System's COSPAS/SARSAT satellites. The latter operates on 406mHz and not only alerts the rescue authorities to your plight but will also give your vessel's identity and nationality. In neither case is voice transmission possible.

For the offshore sailor this is an important piece of kit which (in the case of the cheaper type) can be purchased for about the cost of a comprehensive flare pack. There are many sailors who owe their lives to an EPIRB. It should be high on your list of safety equipment.

Satellite Communications

In this field we already have the future in sight. For virtually global coverage (both poles are excluded) we currently have in place Inmarsat-B (for voice and text) and Inmarsat-C (text only). The former of these two systems uses an ungainly gyro-stabilised antenna, which limits its use to larger vessels, but the latter only requires a compact omni-directional antenna which could be carried by almost any yacht. The text facility works rather like Telex, with a personal computer as a terminal.

Inmarsat-C is perhaps the most versatile current choice for the voyaging businessman. The equipment costs about twice that of Marine Telex, and is some three times more expensive to use. But, it is rather more reliable and less subject to the kind of radio interference to which any HF transmission is prone.

So far as two-way communications are concerned, this is pretty much the state of play as of now. But developments move on apace and – with the exception of VHF, and for the smaller yacht at least – the whole kit and caboodle could be kicked into touch by the arrival of international cellphones. A comprehensive new system of low-orbit satellites are planned, which will provide portable-phone type coverage from almost anywhere. The scheduling of such ambitious projects is always subject to slippage, but by 1997 the first stages should be operational, with the entire network in place some short time after. Quite what all this will do for the peace of mind of the get-away-from-it-all cruising sailor is open to question, but there will certainly be no excuses for not keeping in touch after then.

But not all useful radio communications require a two-way facility. Most sailors already listen to weather forecasts broadcast from domestic radio stations or from coastguard transmissions. And there are other ways of obtaining vital information on navigational matters, though these require dedicated equipment.

Weather Facsimile

There are about ninety stations world-wide which regularly broadcast meteorological maps on HF SSB frequencies. Assuming you have the necessary equipment, and depending upon which system you have chosen, the information can then either be printed out or displayed on a personal computer. There are different ways to achieve this:

- Dedicated Weather Fax Receivers: These are stand-alone units using internal receivers, pre-programmed to tune into the appropriate frequencies. Some will even scan through the frequencies, choosing automatically the one with the strongest signals. The very best of these units are extremely robust, well proofed against the rigours of a marine life, produce high quality print-outs of received information. They tend to be rather expensive, though. Cheaper equipment is to be had, but these can be unreliable.

- Software Based Systems – such as PC Weatherfax. These use IBM PC compatible computers – usually, 'laptops' or 'notebooks', for obvious reasons of convenience – hooked up to an HF SSB receiver. Although it is possible to print out the data (assuming, of course, you have a printer), the principal intention here is to view the computer display itself, thus taking advantage of the superior picture quality. If you already have a computer and an SSB receiver (the latter almost essential for the long-distance sailor, anyway), then buying the software would give you the weather fax facility at relatively little extra cost. Computers are tough little beasts, if treated half-decently, and can perform many other useful tricks for the mariner. Incidentally, weather fax transmissions are in black and white, so for this purpose at least you can avoid the expense of a colour screen.

Communications

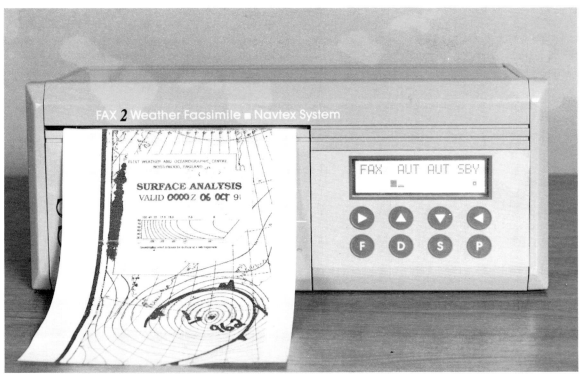

The ICS Fax 2 processes and prints out weather facsimile and Navtex information received by any HF SSB receiver.

   A related but fundamentally different approach is to receive images from the weather satellites themselves. For this you will need a special receiver capable of tuning to the satellites' VHF frequencies, a computer, and the software needed to decode and display the images. These images are derived from photographs in both visible and infra-red light and are transmitted in 'grey scales', which can be translated into some representation of colour by the software.

Navtex

This service is an element of the Global Maritime Distress and Safety System and provides general marine safety information to all vessels within about 300 to 400 miles offshore. Broadcasts are made in many parts of the world and give text presented weather forecasts, ice reports, navigational warnings, and search and rescue alerts. Navtex is transmitted on 518 kHz in English and 490kHz in the local language. It is received on dedicated equipment (though some weather fax receivers can

The Innovative Yacht

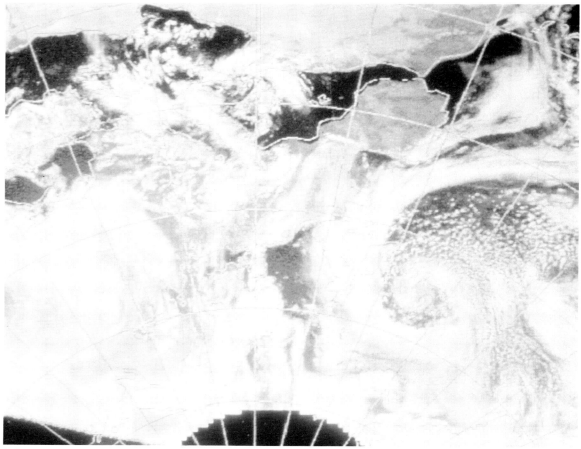

PC Weatherfax software allows weather and Navtex information received by short wave radio to be displayed on a shipboard computer.

also handle Navtex) which either prints it out for you, or stores it within the receiver's memory for later viewing on a CRT screen.

On commercial vessels, Navtex receivers are left on permanently, ready to receive whatever comes their way. But sailboats, with their limited electrical resources, usually need to be more selective. The equipment can be programmed to record only specified categories of information, and may be switched on as and when required.

For the English speaking sailor cruising abroad, Navtex can be an enormous boon. In the Mediterranean, for example, weather forecasts are invariably in the local language and often rattled out at great speed. With the horizon darkening to windward, it is too late to start brushing up your Spanish, or Italian, or Greek.

Communications

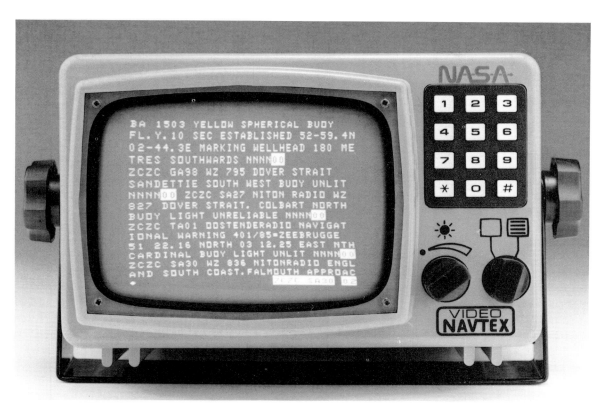

The NASA Marine Navtex receiver is entirely self-contained and displays English language messages on a continuous basis. Up to sixteen pages of information can be stored for review at any time.

Chapter 15
Clothing

Man is ill equipped to face the elements unprotected. In that sense evolution has treated him badly. Although comfortable at home in his prehistoric equatorial African backyard, once he took to wandering he soon found that his naked physique was not up to the harsher demands of other regions. But, being by nature both ingenious and piratical, he quickly observed that the local animals managed just fine in their fur coats; so he killed them and stole the coats from them, and wrapped himself up against the cold. Later he found that the fleece from long haired animals could be twisted into yarns and woven into cloths – lighter and more versatile than heavy hides – and later still that there were other sources of fibres and filaments (flax, cotton, silk) that could be also be utilised.

Clothing has liberated man from geographic limitations. Born, perhaps the most climatically vulnerable, sheer inventiveness has placed him above other animals in terms of adaptability – which he achieves almost in an instant. Not for him the long wait for natural selection to work its evolutionary wonders. He can share the habitat of both arctic fox and desert camel – and more recently the man in the moon! – simply by changing his apparel.

There are, of course, ideal conditions when to be naked or near-naked is to be perfectly attired but, in my experience, sailors are rarely so fortunate. More often one is either too hot, too cold, or too wet, and the transition from one state to the other can happen very quickly. To deal with these changes we need clothing especially developed for the job.

And develop they have. When I first went to sea in the late fifties our uniforms were of 'doeskin' – an expensive, closely woven woollen cloth, about as yielding as roofing felt – with itchy serge trousers and heavy wool sweaters doing duty for rougher work. When it was wet we wore rubber seaboots and oilskins of resin-impregnated canvas which crackled as we moved. Then in the tropics we changed into white cotton duck, starched to a plywood rigidity, and nattily set off with canvas shoes that required daily applications of Blanco to keep them pristine. Thus equipped, we were ready to face all weathers.

Granted this was big ship stuff, but yachtsmen fared little better. I recall an oiled-wool turtle-neck, bought from a fishermen's co-operative, which, the moment it got wet, would sag to my ankles like some grotesque and mightily heavy shroud. And a padded waterproof jacket I had specially made for a long distance race, which had me waddling round the boat like a promotional figure for Michelin tyres – far too physically restricted to be a useful hand on deck. And another, lighter jacket with seams that let in water and a hood so rigid I had to rotate my whole body to see anywhere but straight ahead.

At the time we were grateful for what comfort these relatively primitive garments provided but, happily, things have come a long way since then. Outdoor activities have become big business, and many millions of pounds have been spent in the design and development of the various accoutrements associated with these pursuits – of which clothing forms a major part. And, along with this research has come a new understanding of how we can stay comfortable.

Stay Dry, Stay Warm

Water conducts heat twenty-five times faster than air. Clothing retards this somewhat but the effective thermal conductivity of a typical assembly of wet clothing is still about ten times greater than when dry. And, when water evaporates, it also sucks thermal energy from its surroundings as the moisture is converted into vapour.

When working in such clothes even at an otherwise comfortable ambient temperature of 20°C, the oxygen demands of a person can increase to some fifty per cent in excess of the same work performed in dry gear. Unless the subject is quite fit, this excess can be hard to ventilate for long periods. Consequently, if there is a shortfall in oxygen, the body temperature will fall and there will be an associated risk of exposure. At lower ambient temperatures this threshold will be reached much sooner, and any personal respiratory complaints – such as the common cold – can make matters much worse. This is an insidious process in which a seemingly benign situation can turn very dangerous, almost without anyone noticing.

And here we are not simply talking about bodily survival. With the onset of chilling, there is a serious reduction in both morale and mental acuity. Judgement quickly becomes impaired. Normally sensible sailors might make stupid or desperate decisions quite out of character in other circumstances.

I once sailed with a man who thought it manly to get wet. There he would sit, clad only in his street gear, scoffing at us wimpish individuals huddled in our oilskins. But he finally got his comeuppance after one hard thrash to Alderney. He had started boisterously enough but, as the day wore on, our man became lethargic and morose. His speech became slurred – a classic symptom of exposure – and we pleaded with him to go below. At last in port, and with the rest of the crew getting ready to surge ashore for a well-deserved pint or two, he rose to his feet, ashen of face, and sheepishly admitted he was feeling rough. Seconds later, he pitched forward in a dead faint. A few hours in a warm sleeping bag soon put him to rights but I doubt if it was as enjoyable as the time we other crew members spent touring the pubs. Frankly, he was a fool and a liability to his mates. Such people have no place on any offshore yacht.

In cooler climates it is vital to stay as dry as possible, and for that we obviously need waterproof clothing. High quality wet-weather gear is usually made of a tightly woven nylon cloth coated with polyurethane or neoprene to make it impervious to water ingress. The seams are internally taped to prevent seepage through the stitches, and there are elasticated seals around the wrists and ankles.

Unfortunately, fabrics that keep moisture out will also keep moisture in. Even if comfortably below the sweat threshold, the human body gives off water vapour which can condense on the cooler surface of your oilskins. Soon you can be as damp inside your clothes as you would be if perched naked on the rail. In order to overcome this problem, manufacturers have developed microporous fabrics – so called 'breathable' materials – which allow air and water vapour to pass through but will bar the passage of solid water. However, there is obviously a fine line between making a fabric vapour-permeable and having it leak like a sieve. Sailing often demands bursts of high activity, causing sweating, followed by long static periods when the body can become chilled. No breathable fabric can cope with such extremes on its own and, if we are to remain comfortable under all conditions, we must do more to handle variations in perspiration output.

The design and manufacture of modern sailing clothing is now very reliant upon technology. The emphasis is now on modular systems rather than monumental, do-it-all garments. Whereas, a few years ago you might have bought a set of oilskins which purported to satisfy every need, now you are more likely to choose a layered assembly which you can adjust to suit specific circumstances.

Before we move on to the systems themselves, first a word about the materials. Natural fibres have been virtually abandoned in favour of synthetics. Polyester, polypropylene and polyamide (of which nylon is one) filaments are spun into variously constructed yarns before being woven or knitted into fabrics. Of these, perhaps the most remarkable is Tactel – a polyamide material, developed by ICI and now made by Dupont, which is so strong it can be drawn into strands sixty-times finer than a human hair. Fluffed into a fleece, it has excellent moisture transportation and air-trapping properties; woven tightly it is wonderfully wind and water resistant.

In order to be effective, any system of marine clothing must perform three vital functions:

1. Perspiration must be wicked away from the skin to where it can evaporate without pulling heat directly from the skin's surface. By reducing the sensation of clamminess, this also makes you feel much more comfortable.

2. A stable layer of air must be trapped against the body to provide insulation. Some exchange is necessary to ventilate the skin, but there should be no rapid or uneven displacement of warm air by cold coming in from outside

3. Rain and the inevitable dollops of sea-water spray must be kept at bay.

The first two of these functions can often be achieved in the same fabric – or perhaps a composite of two fabrics bound together. A fine fibre-pile draws the perspiration away from the skin by capillary action. When it reaches the outer layer, the moisture spreads laterally along its woven or knitted surface and simply evaporates into the wind. Any spray or rain water that penetrates to the skin is transported back out the same way. The pile, whose volume is made up largely of voids, also serves to retain the air which then quickly warms to body heat.

This principle is applied to the splendid Sub-Zero range, of which I have blissful experience. Here the fabric is made up of fleeced Tactel, and the basic kit comprises salopettes and a zip-fronted jacket, under which you have underwear – again Tactel – of long-johns and polo-necked shirt. This combination is astonishingly warm and is virtually windproof. Even light rain is disdainfully shrugged off. Thus attired, with some oilskins on hand for the really heavy stuff, you are kitted out to face anything.

A slightly different approach is applied to the Trax system by Remploy, which has received high praise from numerous competitors in circumnavigational and other epic long-distance races. Here, the fabric is made up of a polyester pile, bonded to a special polyamide based outer skin (trade name 'Pertex') which is hard-wearing and virtually windproof. A shirt and salopettes are worn directly against the skin. No underwear is required – indeed, it detracts from the efficacy of the system. As the going gets harsher, the basic outfit can be augmented (or the shirt replaced) by an even warmer jacket. Accessories such as boot liners, hood, and mitts complete the assembly and can be utilised as required.

But, without challenging the thermal qualities of this last system, I have some personal reservations about not wearing underwear. It may be fine for a day or two, but I think the gear could get pretty ripe over longer periods. Thermal underwear is light, easy to wash, and relatively cheap. To carry a couple of sets in the interests of hygiene seems to me the better way to go.

Yet, whichever way you choose, there are enormous advantages to the yachtsman in having specialised clothing. No stuffed crew bags filled with heavy sweaters need be carried – just the shipboard garments and the glad rags for tripping ashore. And no longer those mounds of soggy clothing below, which seem to invade the interior like smelly, sick animals seeking somewhere warm to die.

Crew efficiency and safety is also improved. To those hitherto accustomed to being long-johned and woolied almost to the point of immobilisation, the liberation from these strictures can bring amazement. Comfort is increased. Fatigue is reduced. And that extra nimbleness on deck might even save your life.

The Innovative Yacht

The Trax thermal clothing system by Remploy is well regarded by many long-distance sailors.

The Sub Zero gear is exceptionally warm and comfortable to wear. By donning or discarding layers, you can adjust for different temperature conditions.

Eventually, of course, the weather will become so awful that even greater protection is needed. For this we need wet-weather gear – 'oilies', 'foulies', 'weathers', 'slickers', whatever you care to call them. Either way, this is the sailors' battle armour and its construction owes more to engineering than it does to conventional tailoring.

Serious wet-weather gear – that is, excluding the lighter stuff you might use in a tropical shower – must be very well made to withstand the rigours of its intended environment. Manufacturers tend to signal the relative efficacy (and cost) of their products in terms of mounting heroicisms – 'Inshore', 'Coastal', 'Ocean' being representative. Of course, to spread the price range is to widen the market but, personally, I have found the water to be just as unpleasantly wet half a mile offshore as it is off Muckle Flugga or in the middle of the Atlantic. You get the protection you pay for, and in my opinion within reason you should try to obtain the best.

And the best are very good indeed – infinitely better than was available just a few years ago. Unlike the developments in 'thermal' clothing, the improvements have not been so dependent upon the arrival of new materials, but have been gained by a better understanding of how such gear should be put together.

The most common and versatile arrangement – the classic wet-weather gear – is to wear high-fitting trousers under a thigh-length jacket. The trousers are supported by shoulder straps and should reach to just under the armpits. When worn in combination with the jacket, this gives an overlap over almost the entire torso, and makes it very difficult for water to enter via that route. Wrists are sealed with elasticated or 'Velcro' cuffs and the trouser legs are worn (of course) outside your deck boots, again providing an effective baffle. The whole of this natty ensemble is topped off with a hood which can be either a permanent part of the jacket or fixed with a zip, press-studs or Velcro.

The trousers will get the most use. Decks are often wet from condensation or the aftermath of rain, long after there is a need to wear a waterproof jacket. I find I wear out at least a couple of pairs before my jacket needs replacing. Luckily these items are usually bought separately so, apart from the pain in my wallet, there is no problem here, and it does give you the opportunity to mix-and-match – to choose each garment as it suits you, perhaps from different manufacturers.

Keeping Cool

This is rarely a problem for the British yachtsman, but those of an optimistic turn of mind hope to encounter it one day.

Actually, there is not too much to be said about keeping cool. The usual accoutrements of sunglasses, hats and the correct Sun Protection Factor embalming lotions are familiar accessories to any holiday maker, and are as essential on a sailboat as they are on the beach.

But there are a couple of items of clothing which I heartily recommend. The first is a pair of cotton pyjamas – personally I favour a rather natty paisley pattern here – which make excellent attire for those scorching days where to be naked or nearly so is likely to be barbecued.

The second I was introduced to in Burma, where it is called a longhi – or, to us more familiarly, a sarong. This is, of course, the ubiquitous garment of the Far East, worn by millions of men and women on a daily basis. Constructionally, it could not be simpler: A tube of light cotton fabric, about two metres (6.5ft) in circumference and about one metre (3.28ft) long – the sort of thing that could be run-up on a sewing machine in a matter of minutes. Tied around the waist or – for the ladies – under the armpits, it is wonderfully cool and comfortable. And it is an absolutely guaranteed conversation stopper when, thus clad, you march your crew into the yacht club.

Index

acoustic insulation, 136
acrylic cloth, 74
activated carbon filter, 115
air conditioning 129
alternator, 145
amateur (ham) radio, 172
ampere, 140
anchor, 64
apparent wind, 79
asymmetric spinnaker, 53
autopilot, 86
automatic radar plotting aid, 170
auxiliary rudder, 81

ballast, 29
  -water, 31
  -ratio, 31
barbecue, 104
batten, 50
battery, 143
  -charger, 151
beam, 17, 29
bimini, 73
boom, 55
bottlescrew, 56
bunk, 94
butane, 100

cabin heater, 125
calorifier, 119
capsize, 30
centre of gravity, 29
chain plate, 56
Channel Handicap, 12, 20
chart plotter, 167
coaming, 58
cockpit, 57
  -drain, 59
communication equipment, 171
compressed natural gas, 101
compressor, 108
  -system, 105
condenser–fridge, 108
constant velocity joint, 134
cooker, 100
crimp, 36
cross track error, 167
cruising chute, 53
cutter, 40

Decca, 163
deck, 13, 57
  -awning, 74
  -saloon, 70
depth-sounder, 158

desalinator, 115
diesel engine, 130
differential GPS, 165
discharge treatment system, 123
displacement, 17, 20, 24
   -heavy, 25
   -light, 27
   -medium, 26
   -ultra-light, 28
Displacement/Length Ratio, 20
dodger, 72
dog house, 71
drag, 34
drain, 59
draught, 17
drinking water, 115

electrical equipment, 139
electronic autopilot, 86
   -log, 160
Emergency Position Indicating Radio Beacon, 173
engine, 14, 130
engine-noise, 132
   -size, 130
eutectic plate, 108
exhaust system, 135

ferro-resonant charger, 152
fiddle, 103
flexible shaft coupling, 133
fluxgate-compass, 86, 161
forward drive component, 33
fractional rig, 40
fuel filter, 138
full-length batten, 50

galley, 98
gas, 101
genoa roller reef, 46
Global Maritime Distress System, 173
Global Positioning System (GPS), 164

goose neck, 56
grab-rail, 61
ground-tackle, 64
guardwire, 61
GZ curve, 30

heads, 122
headsail, 36
heat exchanger, 120
heeling component, 33
holding tank, 122
hull, 13
   -resistance, 16
   -speed, 19

icebox, 109
in-boom mainsail roller reef, 48
in-mast mainsail roller reef, 46
Inmarsat-B, 173
   -C, 173
insect screen, 77
International Measurement System, 12, 20
International Offshore Rule, 12, 19

jackstay, 61

keel-long, 25
ketch, 43
kicking strap, 56

laminated sail, 38
lazy jack, 51
lee cloth, 95
length, 17
lift, 34
lift/drag ratio, 35
Liquid Petroleum Gas (LPG), 100
log, 160
Loran-C, 162

Index

mainsail, 36
marine telex, 173
mast, 55
   -unstayed, 44
masthead, 56
   -rig, 39
mechanical similitude, 17
Mediterranean moor, 67
microwave oven, 104
monohull, 16
multihull, 16

Navtex, 175
NMEA, 158, 166
non-slip surface, 61

passarelle, 66
pendulum-servo gear, 83
plane, 19
power consumption, 141
pressure cooker, 105
prismatic coefficient, 22, 24
propane, 100
pulpit, 62, 67

radar, 169
radio, 171
reefing system, 46
refrigeration, 105, 112
regulator, 146
resistance–hull, 16
rig, 13, 33
rigging screw, 56
righting arm, 29
rudder, 83
   -auxiliary, 81

sacrificial strip, 46
sail, 33
   -area, 17, 22, 24
sail area/displacement ratio, 22
sail plan, 13, 38

sailcloth, 35
SatNav, 164
sea berth, 95
seacock, 59
self-steering, 78
self-tacking staysail, 41
self-tailing winch, 62
shower, 121
shroud, 55
silencer, 135
silicon controlled rectifier, 152
single line reef, 49
single side-band (SSB) radio, 172
skin drag, 18
slab reef, 48
sloop, 39
solar panel, 149
Speed/Length Ratio, 17
spinnaker, 51
   -pole, 55
spray hood, 71, 72
spreader, 56
stability, 17, 25, 29
stanchion, 62
staysail, 40
storm jib, 54
   -trysail, 54
tang, 55
telex, 173
thermo-electric refrigeration, 110
tiller, 60
toggle, 56
Transit, 164
trim-tab, 82
turnbuckle, 56

ultra-violet light, 38

velocity made good indicator, 161
VHF, 171
vibration, 132

volt, 140
voltage controller, 152

warp, 35, 64
water ballast, 31
   -filter, 115
   -heater, 119
water-lift silencer, 135
waterline length, 24
watt, 140
wave-making resistance, 18
waypoint, 166
weather facsimile, 174

weft, 35
wet-weather gear, 183
wetted surface area, 17, 24
wheel steering, 60
wheelhouse, 69
winch, 62
wind generator, 147
   -indicator, 161
windlass, 64
windscoop, 76
windscreen, 72
windvane gear, 79